Fundamentals of Microbiology, Tenth Edition

Jeffrey C. Pommerville, PhD

1914–2014
Nelson Education celebrates 100 years of Canadian publishing

NELSON EDUCATION
CELEBRATE LIFELONG LEARNING

NELSON EDUCATION

Printed and bound in Canada
1 2 3 4 17 16 15 14

For more information contact Nelson Education Ltd., 1120 Birchmount Road, Toronto, Ontario, M1K 5G4. Or you can visit our Internet site at http://www.nelson.com

This textbook is a Nelson custom publication. Because your instructor has chosen to produce a custom publication, you pay only for material that you will use in your course.

ISBN-13: 978-0-17-659781-8
ISBN-10: 0-17-659781-6
Consists of Selections from:

Fundamentals of Microbiology, Tenth Edition
Jeffrey C. Pommerville, PhD
ISBN 10: 1-4496-8861-6, © 2013

Cover Credit:
Sven Hoppe/Shutterstock

Brief Contents

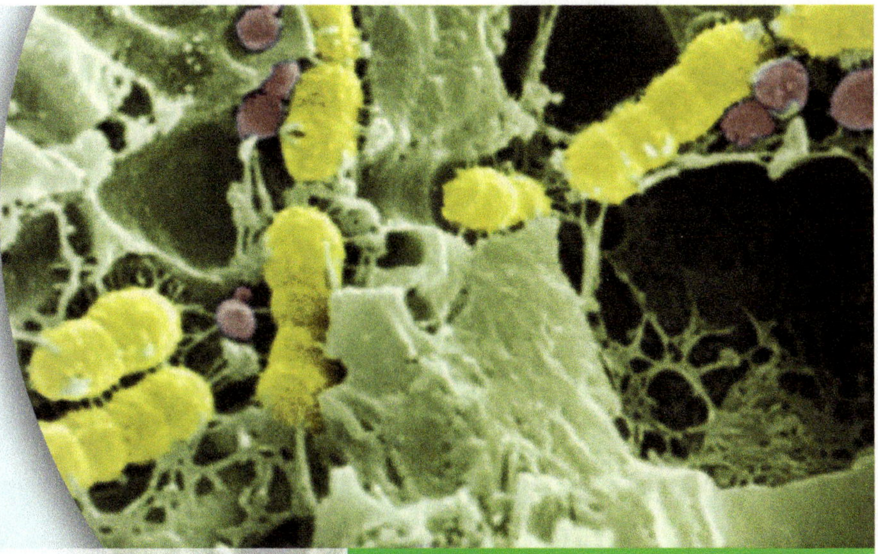

Gouda cheese consists of a casein (milk protein) matrix interspersed with fat globules and lactic acid bacteria. The bacterial cells shown in yellow and pink are streptococci or lactococci. © Scimat/Photo Researchers, Inc.

Environmental and Applied Microbiology

PART 6

Public health has many facets. Several—such as mental health, nutritional health, and environmental health—lie outside the realm of this text. Other facets, however, are concerned with communicable disease and therefore fall within the scope of microbiology. Immunization, for example, is a public health measure because it creates a barrier within the individual to lessen susceptibility to infection. Antibiotic treatments are also under the umbrella of public health because they are used to limit the spread of communicable disease.

In Part 6, we are concerned with sanitation, a facet of public health designed to prevent microorganisms from reaching the body. Sanitation came into full flower in the mid-1800s, and it is a relatively recent phenomenon. Before that time, living conditions in some Western European and American cities were almost indescribably grim: Garbage and dead animals littered the streets; human feces and sewage stagnated in open sewers; rivers were used for washing, drinking, and excreting; and filth was rampant.

The belief that filth was a catalyst to disease fueled the sanitary movement of the mid-1800s (see Chapter 1). As the Industrial Revolution sparked great population increases in the cities, health problems mounted, and sanitary reformers spoke up for effective sewage treatment, water purification, and food preservation. The germ theory of disease, developed in the 1870s, strengthened the movement and justified its actions because communicable disease could now be blamed on microorganisms in food and water. As sanitary reformers merged with bacteriologists, they issued loud calls for government support of public health and stimulated the development of new methods for dealing with food and water, which we study in Chapters 25 and 26.

As you know, only a very small minority of microorganisms are pathogens. In fact, some contribute mightily to public health. For example, certain microorganisms are used to produce a broad assortment of foods and dairy products making up a regular part of our diet. In addition, microorganisms play a dominant role in numerous cycles of elements in the environment and forge a link between what is useless and what is useful to other living organisms. Moreover, scientists employ microorganisms on industrial scales to synthesize a variety of products we could not obtain otherwise. We see examples of these contributions in Chapters 25, 26, and 27.

MICROBIOLOGY PATHWAYS

Public Health Inspector

© Shout/Alamy Images.

How often have you seen on the television or in the newspaper a report on "dirty restaurants" where there is "slime in the ice machine" or "meat kept at inappropriate temperatures"? These reports are the result of visits by state, county, or city food or public health inspectors.

Nationally, the U.S. Department of Agriculture's Food Safety and Inspection Service (FSIS) employs more than 7,500 inspectors who comprise an essential component of the nation's public health mission. The food inspectors in private commercial slaughtering plants provide the expertise to ensure diseased and contaminated meat and poultry does not enter the food supply. Therefore, they are responsible for most of a slaughtering plant's daily inspection of animals—both pre- and post-slaughter.

One career path for a food inspector is a consumer safety inspector in private meat, poultry, or egg processing plants. Such inspectors ensure the plant is operating within approved food safety guidelines.

Another career path is as an import inspector. Import inspectors work at seaports and other points of entry into the United States. They make sure products imported from other countries are as safe as those produced domestically.

To qualify for an entry-level position, you must pass a written test and have either a bachelor's degree or one year of job-related experience in the food industry. This experience must demonstrate knowledge of sanitation practices and control measures used in the commercial handling and preparation of food products for human consumption.

Another career path is as a public health inspector, sometimes called an environmental health officer. These individuals help prevent infectious disease and promote health by educating people about public health issues and enforcing health legislation. They are the ones looking for the "slime in the ice machine" and other dirty restaurant violations. Inspectors monitor conditions in restaurants and schools, food processing plants, hotels and motels, child care and long-term care facilities, animal facilities, swimming pools, and recreational camps. Public health inspectors also may investigate reportable diseases.

To become a public health inspector you must have a community college or university degree in an appropriate specialization such as public and environmental health, food sciences, or water chemistry. You must have a strong interest in the applied sciences and an interest in community service; be curious, decisive, observant; and have good judgment and excellent interpersonal, investigative, and communication skills. And of course, having courses and a background in microbiology is a real bonus!

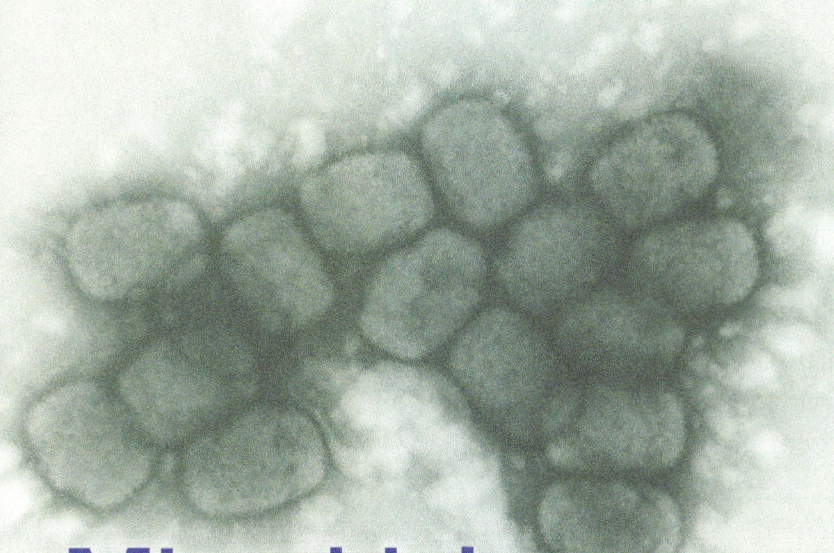

Microbiology of Foods

We are not losing the "war." Unfortunately, we probably can't win it either. Food safety folks will always have a job.
—Arthur P. Liang, MD, MPH, Centers for Disease Control and Prevention

The idea was appealing and the price was right: a patty melt sandwich and a soft drink for lunch. The rye bread was toasted, the hamburgers were stacked and waiting to be cooked, the American cheese slices were lined up next to the grill, and the aroma from the sautéed onions was irresistible. It was October 1983, at the Skewer Inn, a restaurant at the Northwoods Mall in Peoria, Illinois. The stage was set for one of the worst outbreaks of botulism in American history.

Between October 14 and 16, many people stopped by the restaurant and enjoyed patty melt sandwiches. Soon, however, 36 individuals began experiencing the paralyzing signs of botulism. They suffered blurred vision, difficulty swallowing and chewing, and labored breathing. One by one they called their doctors, and within a week, all were hospitalized. Twelve patients had to be placed on respirators. After many anxious hours, all but one recovered.

Investigators from the Centers for Disease Control and Prevention (CDC) arrived in Peoria shortly thereafter. They obtained detailed food histories from patients and from others who ate at the restaurant during the same 3-day period. First, they identified patty melt sandwiches as the probable cause (24 of 28 patients interviewed specifically recalled eating the sandwiches); then they began a search to pinpoint the item that might be responsible. The data pointed to the onions. Investigators isolated *Clostridium botulinum* spores from the skins of fresh onions at the restaurant, and learned that after sautéing, the onions were left uncovered on the warm stove for hours. Furthermore, the onions were not reheated before serving. Spores had probably germinated within anaerobic mounds of warm onions and the vegetative cells deposited their deadly toxins.

25

CHAPTER PREVIEW

Image courtesy of Dr. Fred Murphy/CDC.

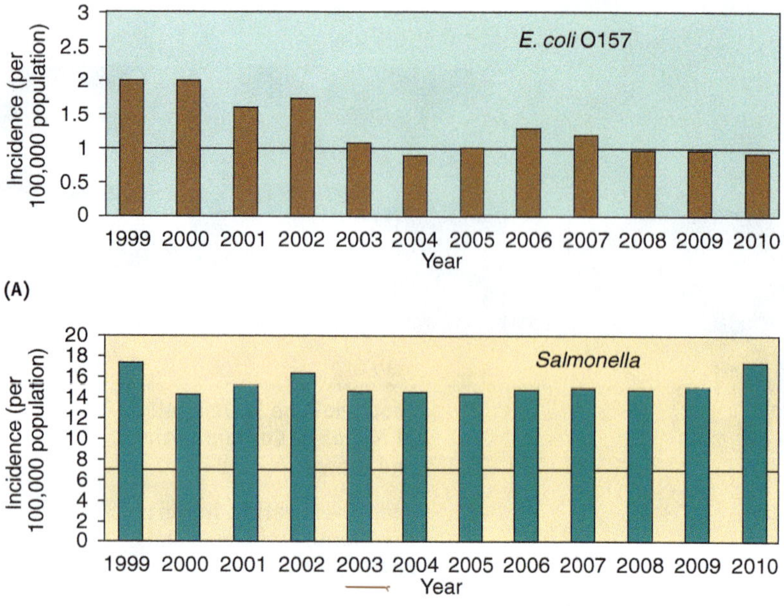

FIGURE 25.1 Incidence of Selected Foodborne Illnesses—United States, from 1999–2010. The incidence of foodborne illness caused by **(A)** *E. coli* 0157:H7 and **(B)** *Salmonella*. The national health objective is to reach a level where the incidence is no more than 1 per 100,000 people for *E. coli* and 6.8 per 100,000 for *Salmonella*. *Source:* FoodNet/CDC. »» Provide some ways in which the incidence of foodborne diseases caused by *E. coli* and *Salmonella* can be reduced.

Incidents like this one occur too often in the United States. The CDC estimates more than 48 million Americans get sick (1 person in every 6), over 128,000 are hospitalized, and 3,000 people die each year from foodborne illnesses. However, efforts by the meat industry are attempting to reduce the number of foodborne illnesses (FIGURE 25.1). For example, *Escherichia coli* O157:H7 infections declined 36% between 2002 and 2003, the largest decline ever. Since 1998, *E. coli* O157:H7 infections have declined over 40%, while others, such as those caused by *Salmonella* species, have remained relatively stable.

The national health objective is to reach a level where the incidence of these and other foodborne illnesses, such as *Campylobacter* and *Listeria*, is no more than 1 per 100,000 people. However, *E. coli* (STEC) infections transmitted through produce have been increasing. The 2010 outbreak of *E. coli* O104:H4 in fresh bean sprouts in Europe is but one example.

The number of Americans infected every year illustrates how most foods, even cooked foods, provide excellent conditions for the growth of microorganisms and the depositing of their toxins. The organic matter in food is plentiful, the water content is usually sufficient, and the pH is either neutral or only slightly acidic. To the food manufacturer or restaurant owner, the growth of microorganisms may spell economic loss or a sullied business reputation. To the consumer, it may mean illness or, in some cases, death.

The Foodborne Diseases Active Surveillance Network (FoodNet) within the CDC is a collaborative project administered at ten U.S. sites along with the Department of Agriculture and the Food and Drug Administration (FDA). FoodNet actively surveys for foodborne diseases and carries out epidemiologic studies to help public health officials better understand the epidemiology of foodborne diseases in the United States.

The primary focus of this chapter is to examine the types of microorganisms that contaminate various foods and to point out the consequences of contamination. We examine food spoilage and the preservation methods used to prevent spoilage. Toward the end of the chapter, we describe some forms of microbial growth in food that are actually desirable because they generate the numerous fermented foods we consume regularly.

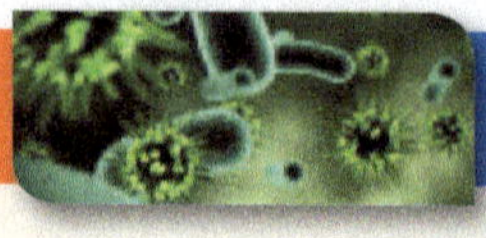

Chapter Challenge

In many parts of the world, humans continue to consume milk beyond infancy. In fact, there are more than 6 billion consumers of milk and milk products, the majority of them in the developing world. Milk from cows is an important nutrient, containing significant amounts of saturated fat, protein, and calcium as well as a source for many vitamins. In developed countries, such as the United States, how is milk processed once it is harvested from the milk cows and sent to the dairy or factory site for processing? How does milk spoil, what is done to prevent spoilage, and, besides being a beverage, what are some of the fermentation products produced from milk? Let's find out as we learn about food microbiology.

KEY CONCEPT 25.1 Food Spoilage Is Generally a Result of Microbial Contamination and Growth

Food spoilage has been a continuing problem ever since humans first discovered they could produce more food than they could eat in a single meal. We all know of Marco Polo's travels to China in the 13th century to obtain spices and explore new trade routes. What often is not mentioned is that the spices were more than just a luxury of the time. Spices were essential for improving the smell and taste of spoiled food. Refrigeration was virtually unknown, and canning was yet to be invented.

Food Spoilage Comes from Many Microbial Sources

Spoiled foods usually have an unpleasant appearance, aroma, and taste. Sometimes, however, these signs may be difficult to detect, such as when staphylococci deposit enterotoxins in food or when too few bacteria grow to cause a perceptible change. The consumption of toxins or microorganisms may cause a number of food poisonings (intoxications) or infections.

Enterotoxin: A bacterial exotoxin that can cause gastroenteritis.

Contaminating microorganisms enter foods from a variety of sources. Airborne organisms, for example, fall onto fruits and vegetables and then penetrate the product through an abrasion of the skin or rind; crops carry soilborne bacterial species to the processing plant; shellfish concentrate organisms by straining contaminated water and catching the organisms in their filtering apparatus; and rodents and arthropods transport microorganisms on their feet and body parts as they move about among foods.

Human handling of foods also provides a source of contamination. For example, bacterial species from an animal's intestine contaminate meat handled carelessly by a butcher. Raw vegetables, such as those typically found at salad bars, also are of great concern for spoilage.

Several Conditions Can Determine if Spoilage Will Occur

Because food is basically a culture medium for microorganisms, the chemical and physical properties in and surrounding the food have a significant bearing on the type of microorganisms growing on or in it (FIGURE 25.2).

Conditions naturally present in foods that influence microbial growth are called **intrinsic factors**. These include:

Water. One of the prerequisites for all life is water. Therefore, food must be moist, with a minimum water content of 18% to 20% before contamination by microorganisms and spoilage can occur. Microorganisms do not grow in foods such as dried beans, rice, and flour because of their low water content.

pH. Another important factor is a food's acidity. Most foods fall into the slightly acidic range on the pH scale, and numerous bacterial species multiply under these conditions. In foods with a pH of 5.0 or below, acid-loving molds often replace the bacterial species. Citrus fruits, for example, generally escape bacterial spoilage but yield to mold contamination.

Physical Structure. Another property of a food is its physical structure. A raw steak, for example, is not likely to spoil quickly because microorganisms cannot penetrate the meat easily. However, raw, ground hamburger meat can deteriorate rapidly because microorganisms exist both within the loosely packed ground meat as well as on the surface.

Chemical Composition. A food's chemical composition (nutrients) may be another intrinsic factor determining the type of spoilage possible. Fruits, for instance, support organisms metabolizing carbohydrates, whereas meats support protein decomposers. Starch-hydrolyzing bacterial cells and molds often are found on potatoes, corn, and rice products. The presence of certain vitamins encourages particular microorganisms to proliferate, while the absence of other vitamins provides natural resistance to decay.

The food industry recognizes three groups of foods, loosely defined on the basis of their intrinsic properties (FIGURE 25.3). **Highly perishable** foods are those that spoil rapidly. They include poultry, eggs, meats, most vegetables and fruits, and dairy products. Foods such as nutmeats, potatoes, and some apples are considered **semiperishable**, meaning they spoil less quickly. Orange juice is usually considered semiperishable because of its high acidity. **Nonperishable** foods are often stored

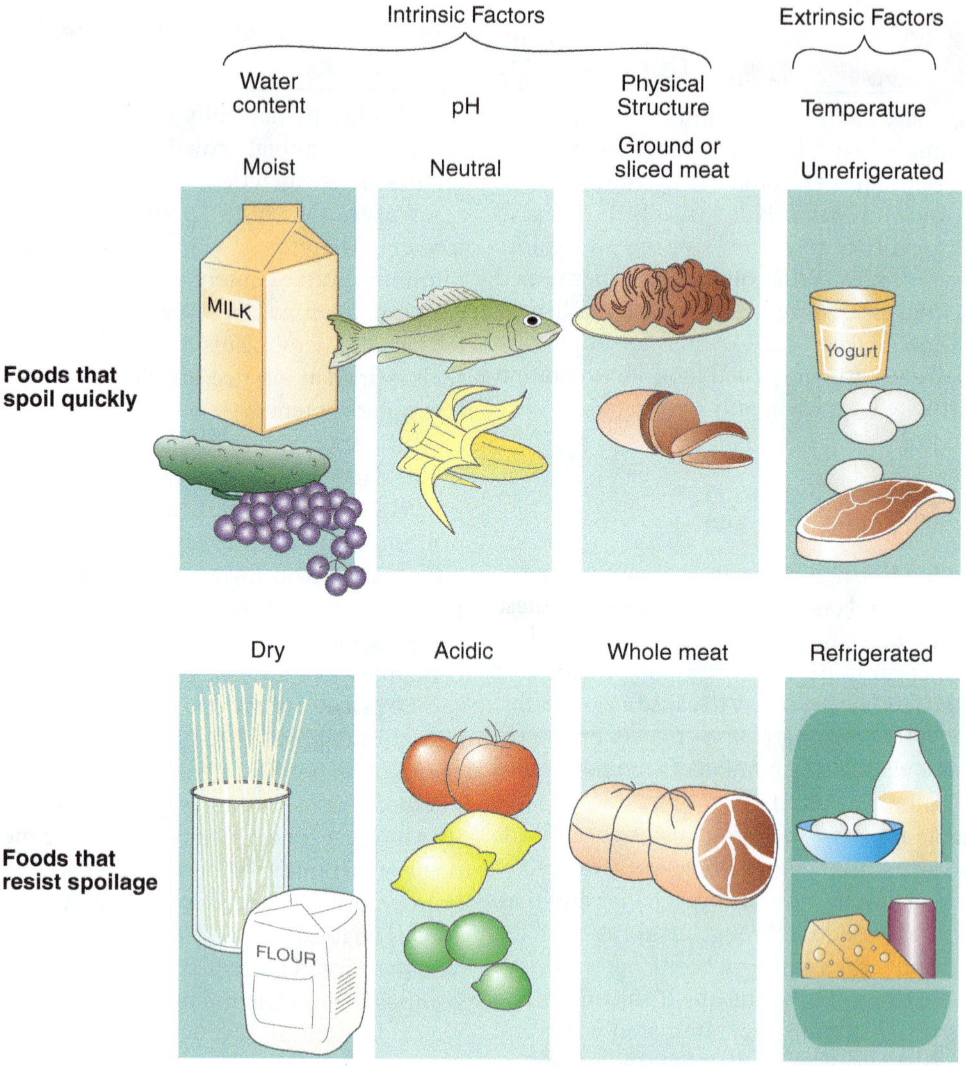

FIGURE 25.2 **Food Spoilage.** Several intrinsic and extrinsic factors determine whether foods are likely to spoil quickly or resist spoilage. »» Why are unrefrigerated foods more likely to spoil than refrigerated ones? Neutral ones versus acidic ones?

FIGURE 25.3 **The Perishability of Foods.** Examples of highly perishable, semiperishable, and nonperishable foods. The physical and chemical properties of these foods are reliable indicators of their rate of perishability. »» How do highly perishable foods differ from semiperishable ones?

in the kitchen pantry. Included in this group are cereals, rice, dried beans, macaroni and pasta products, flour, and sugar.

Environmental conditions surrounding the food (food storage and packaging) make up the so-called **extrinsic factors** influencing microbial growth.

Oxygen. Properly vacuum-sealed cans of food are without oxygen gas and thus do not support the growth of aerobic bacterial pathogens such as *Pseudomonas*. However, an improperly prepared food contaminated with the anaerobe *Clostridium botulinum* provides a suitable environment for growth and one may contract botulism if the food product is consumed. Likewise, vegetables and baked goods do not support the growth of anaerobes.

Temperature. Refrigerator temperature is usually too cold for the growth of most human pathogens (except the psychrotrophic members like *Listeria*) and freezer temperature halts the growth of microbes. However, the warm hold of a ship or a humid, hot warehouse storeroom is an environment conducive to the growth of many human pathogens. It is common knowledge that contamination is more likely in cooked food at a warm temperature than cooked food kept at refrigerator temperature.

The Microorganisms Responsible for Spoilage Produce Specific Products

Often the unpleasant appearance, aroma, and taste of spoiled foods is the result of metabolic products produced by the contaminating microorganisms. Yeasts, for example, live in apple juice and convert the carbohydrate into ethyl alcohol, a product giving spoiled juice an alcoholic taste. Certain bacterial species convert food proteins into amino acids, then break down the amino acids into foul-smelling end products. Cysteine digestion, for example, yields hydrogen sulfide, which imparts a rotten egg smell to food. Tryptophan digestion yields indole and skatole, which give food a fecal odor.

Two other possible products of the microbial metabolism of carbohydrates are acid, which causes food to become sour, and gas, which causes sealed cans to swell. Moreover, when fats break down to fatty acids as in spoiled butter, a rancid odor or taste may evolve. Capsule production by certain bacterial species causes food to become slimy, and pigment production imparts color. In numerous historical incidents, the red pigment from *Serratia marcescens* in bread has been interpreted as a sign of blood (MicroFocus 25.1).

Some foods resist spoilage naturally because they contain antimicrobial chemicals. The white of an egg has lysozyme, an enzyme that digests the cell wall of gram-positive bacteria. Garlic contains certain compounds that inhibit many bacterial species. Oil of clove appears to have healing tendencies, while radish and onion extracts both retard bacterial growth.

Meat and Seafood Can Become Contaminated in Several Ways

Meat and seafood normally are free of contamination because the muscle tissues of living animals normally are **sterile**. Spoilage organisms enter during handling, processing, packaging, and storage. For example, if a piece of meat is ground for hamburger, microorganisms from the surface accumulate in the teeth of the grinder along with other dustborne organisms. Bacterial cells from preparers' hands or from a sneeze compound the problem. The grinder may be cleaned well and refrigerated, but rarely is it sterilized. The importance of foodborne infection in ground meat is pointed up by the recalls of ground beef that occur periodically in the United States.

Sterile: Devoid of living microorganisms, viruses, and spores.

There also is the problem of the so-called "choke points," or places where animals come together and epidemics spread. In the United States, for example, about 9,000 farms produce calves for beef; the animals then are sent to about 46,000 feedlots for developing, and then, to about 80 plants for slaughter. Microorganisms can spread at any of these points. Moreover, animal feeds often contain the entrails of poultry as added protein sources, and thus, another opportunity for pathogen spread is presented.

Processed meats, such as luncheon meats, sausages, and frankfurters, represent special hazards because they are handled often. Also, natural sausage and salami casings made from animal intestines may contain residual bacterial cells, especially botulism spores. When preparers pack such casings tightly with meat, *Clostridium botulinum* may multiply and produce its powerful toxins. As early as the 1820s, people recognized

MICROFOCUS 25.1: History

The Blood of History

Because of its characteristic blood-red pigment and propensity for contaminating bread, *Serratia marcescens* has had a notable place in history. For example, the dark, damp environments of medieval churches provided optimal conditions for growth on sacramental wafers used in Holy Communion. At times, the appearance of "blood" was construed to be a miracle. One such event happened in 1264 when "blood dripped" on a priest's robe. The event was later commemorated by Raphael in his fresco The Mass of Bolsena. Unfortunately, religious fanatics used such episodes to institute persecutions because they believed that heretical acts caused the "blood" to flow.

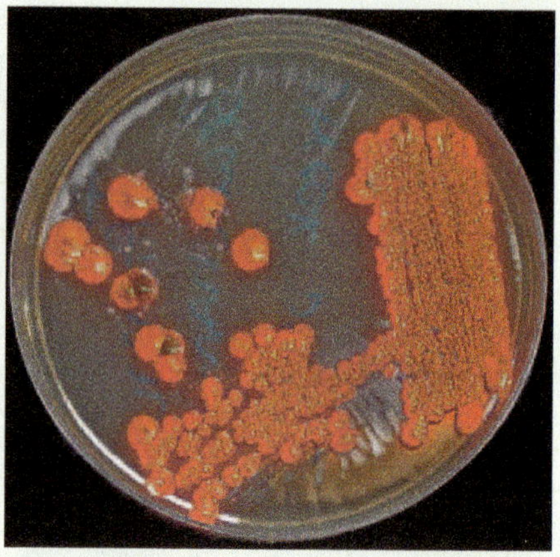

Serratia marcescens colonies growing on agar have the look of blood. Courtesy of Jeffrey Pommerville.

It was not until 1819 that Bartholemeo Bizio, an Italian pharmacist, demonstrated that the bloody miracles were caused by a living organism. Bizio thought the organism was a fungus. He named it *Serratia* after Serafino Serrati, a countryman whom he considered the inventor of the steamboat. The name *marcescens* came from the Latin word for "decaying," a reference to the decaying of bread.

In 332 BC, Alexander the Great and his army of Macedonians laid siege to the city of Tyre in what is now Lebanon. The siege was not going well. Then one morning, blood-red spots appeared on several pieces of bread. At first it was thought to be an evil omen, but a soothsayer named Aristander indicated the "blood" was coming from within the bread. This suggested that blood would be spilled within Tyre and the city would fall. Alexander's troops were buoyed by this interpretation and with renewed confidence they charged headlong into battle and captured the city. The victory opened the Middle East to the Macedonians. Their march did not stop until they reached India.

the symptoms of "sausage poisoning" and coined the name "botulism" (*botulus* = "sausage").

Organ meats, such as livers, kidneys, and sweetbreads (thymus and pancreas), are less compact tissue than muscle and thus spoil more quickly. Moreover, because the organs contain many natural filtering tissues, bacterial cells tend to be trapped in the tissue. Foods like these should therefore be cooked as soon as possible after purchase.

The extent of contamination in meats often consists of a harmless "greening" seen on the surface of a steak. This discoloration is commonly due to the gram-positive rod *Lactobacillus*, or the gram-positive coccus *Leuconostoc*. The green color results from pigment alteration in the meat and represents no hazard to the consumer. Slime formation on the outer casings and souring in frankfurters, bologna, or processed meats also may result from these organisms as well as from *Streptococcus* species.

Microbiologists often can trace spoilage in seafood to the water from which it was taken or held. Fish tissues deteriorate rapidly and the filleting of fish on a bloodstained block encourages contamination. Interestingly, the bacterial species in fish are naturally adapted to the cold environment in which most fish live and thus, cooling will not affect them; freezing is preferred. Shellfish are of particular concern because they commonly obtain their food by filtering particles from the water. Clams, oysters, and mussels concentrate such pathogens as hepatitis A viruses, typhoid bacilli, cholera vibrios, or amoebic cysts. Many cases of cholera have been related to raw oysters.

Poultry and Eggs Can Spoil Quickly

Contamination in poultry and eggs may reflect human contamination, but it usually stems from microorganisms that have infected the bird.

Members of the genus *Salmonella,* with over 2,400 pathogenic serotypes, may cause diseases in chickens and turkeys, then pass them to consumers via poultry and egg products. Processed foods such as chicken pot pies, whole egg custard, mayonnaise, eggnog, and egg salad also may be sources of salmonellosis. Psittacosis is another problem because *Chlamydia psittaci* infects poultry. In one instance, 27 employees in a turkey-processing plant contracted psittacosis while preparing consumer products from diseased turkeys.

Eggs are normally sterile when laid, but the outer waxy membrane, as well as the shell and inner shell membrane, may be penetrated by bacterial species. *Proteus* species cause black rot in eggs as hydrogen sulfide gas accumulates from the breakdown of cysteine. This gives eggs a rotten odor. Other spoilages in eggs include green rot from *Pseudomonas* species and red rot from the growth of *Serratia marcescens.* Green rot results from a fluorescent pigment that causes the egg to glow when placed in front of an ultraviolet light. Red rot gives the yolk a blood-red appearance.

The primary focus of contamination in an egg is the yolk rather than the white. This is because of the nutritious quality of the yolk and because the white has a pH of approximately 9.0. Also, the lysozyme in egg white is inhibitory to gram-positive bacteria. However, *Salmonella enterica* remains a common problem in egg products (FIGURE 25.4).

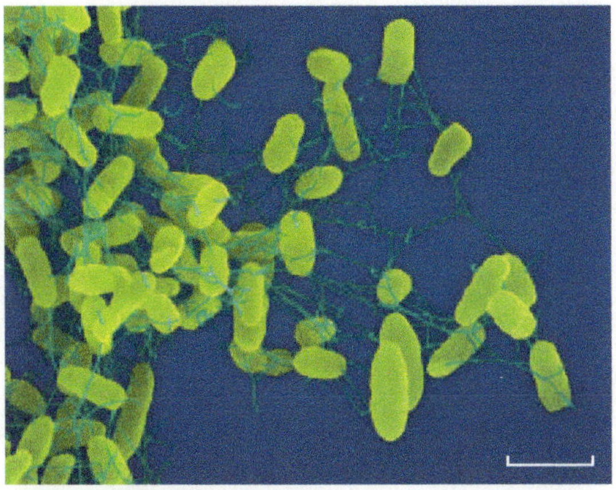

FIGURE 25.4 ***Salmonella enterica.*** *S. enterica* serotype Enteritidis is one of the most widely found contaminants in poultry and egg products. (Bar = 2 µm.) © Dennis Kunkel Microscopy, Inc./Phototake/Alamy Images. »» How might *S. enterica* infect eggs or poultry?

Breads and Bakery Products Can Support Bacterial and Fungal Growth

In the production of bakery products, ingredients such as flour, eggs, and sugar can be a source of spoilage organisms, although most contaminants are killed during baking (MicroFocus 25.2). Some bacterial and mold spores may survive because bread is heated internally to 100°C for only about 9 minutes. Members of the spore-forming genus *Bacillus* commonly survive, and as they proliferate, their capsular material accumulates, giving the bread a soft cheesy texture with long, stringy threads. The bread is said to be "ropy."

Cream fillings and toppings in bakery products provide excellent chemical and physical conditions for bacterial growth. For example, custards made with whole eggs may be contaminated with *Salmonella* species, and whipped cream may contain dairy organisms such as species of *Lactobacillus* and *Streptococcus.* The acid produced by these bacteria results in a sour taste. High sugar environments of chocolate toppings and sweet icings support the growth of fungi. Most bakery products should be refrigerated during warm summer months.

Some Grains Are Susceptible to Spoilage

The most common spoilage microbe in grains is molds due to their ability to survive at low water concentrations. The major concern with mold spoilage is the production of mycotoxins. For example, the ascomycete *Aspergillus flavus* produces **aflatoxins**, a group of toxins that accumulate in stored grains such as wheat as well as peanuts, soybeans, and corn. Scientists have implicated aflatoxins in liver and colon cancers in humans. The toxins are consumed in grain products, as well as meat from animals that feed on contaminated grain.

Another type of grain spoilage is caused by *Claviceps purpurea,* the cause of ergot poisoning (**ergotism**). Rye plants are particularly susceptible to this type of spoilage, but wheat and barley grains may be affected also. The toxins deposited by *C. purpurea* may induce convulsions and hallucinations when consumed.

Milk and Dairy Products Sometimes Spoil

Milk is an extremely nutritious food. It is an aqueous solution of proteins, fats, and carbohydrates containing numerous vitamins and minerals. Milk

MICROFOCUS 25.2: Biotechnology/Environmental Microbiology

Sourdough Bread—Don't Leave San Francisco Without It

Most of us are familiar with the unique taste of sourdough bread. It has a crusty surface and a chewy, but sharply acidic taste. This most distinctive bread, which certainly tastes nothing like typical white sandwich bread, supposedly has its origins with San Francisco. Local stories say the recipe was brought to the city by Basque immigrants from the Pyrenees during the heyday of the California gold rush. So, what gives the bread its distinctive flavor and taste? It is, of course, the resident microbes!

A sourdough baguette. © Jones & Bartlett Learning. Photographed by Kimberly Potvin.

To make sourdough you need a starter culture, which is usually a piece of unused dough from a previous batch. In fact, the San Francisco bakeries say they have kept sourdough cultures going for over 100 years.

Besides the starter, making sourdough only requires unbleached flour, water, and salt. Most sandwich breads have at least 25 ingredients and additives—and no starter culture, meaning yeast is the only microbe present in the rising process for the dough. The result will be relatively bland tasting bread.

Therefore, the most important "ingredients"—and key to sourdough bread—are the microorganisms living in the starter. The sourdough starter is a mixed microbial population of wild yeasts and bacterial species. A teaspoon of sourdough starter contains some 50 million yeasts and 5 billion lactobacilli.

Sourdough is sour because of the acids produced by the lactobacilli. Initially, there may be other bacterial species in the dough, but they are quickly competed out because of the acids produced by the lactobacilli; the pH reaches about 3.8. The bacterial cells also add flavor and produce the characteristic smell of sourdough bread.

So, a sourdough starter represents a mutualistic symbiosis between the principal bacterium *Lactobacillus sanfranciscensis* and the principal yeast, *Candida milleri*, which can tolerate the acidic conditions. So, the next time you pass through the San Francisco airport, be sure to take some *Lactobacillus sanfranciscensis*-laced bread back home—or buy a starter culture and make your own bread at home.

has a pH of about 7.0 and is an excellent nutrient source for humans and animals, as well as microorganisms.

About 87% of the substance of milk is water. Another 2.5% is a protein called **casein**, actually a mixture of three long chains of amino acids suspended in fluid. A second protein in milk is lactalbumin. This protein forms the surface skin when milk is heated to boiling. Lactalbumin is a whey protein, one that remains in the watery fluid after the casein curdles during milk spoilage or fermentation. Carbohydrates make up about 5% of the milk. The major carbohydrate is lactose, sometimes referred to as milk sugar. Rarely found elsewhere, lactose is a disaccharide digested by relatively few bacterial species, and these are usually harmless. The last major component of milk is butterfat, a mixture of fats usually churned into butter. Butterfat comprises about 4% of the milk and is removed in the preparation of nonfat (skim) milk or low-fat milk. When bacterial enzymes digest fats into fatty acids, the milk becomes rancid and is unfit to drink.

■ Whey: The watery liquid that separates from the solid part (casein) of milk when it turns sour or when enzymes are added in cheese making.

A common type of milk spoilage often takes place in the kitchen refrigerator or dairy case at the supermarket. Here, *Lactobacillus* or *Streptococcus* species introduced during milking and processing multiply slowly and ferment the lactose in milk to lactic and acetic acids. Enough acid may develop to change the structure of the protein and cause it to solidify as a curd. Dairy microbiologists refer to such an acidic curd as a **sour curd**. The lactobacilli and streptococci causing it usually have survived the pasteurization process.

Sweet curdling in milk may result when enzymes from species of *Bacillus, Proteus, Micrococcus,* or certain other bacterial genera attack casein. As weak hydrogen bonds break, casein loses its three-dimensional structure and curdles. The reaction is said to be sweet because little acid production occurs. It is an essential step in the production of cheese. The clear liquid is whey, an aqueous solution of lactose, minerals, vitamins, lactalbumin, and other milk components. Whey is used to make processed cheeses and "cheese foods."

Milk also may be contaminated by gram-negative rods of the coliform group, including *E. coli* and *Enterobacter aerogenes*. These bacteria produce acid and gas from lactose. The acid curdles the protein, and the gas forces the curds apart, sometimes so violently that they explode out of the container. The result is therefore called stormy fermentation. *Clostridium* species also cause this reaction.

Ropiness in milk is similar to that in bread. It develops from capsule-producing organisms, such as *Alcaligenes, Klebsiella,* and *Enterobacter*. These gram-negative rods multiply in milk, even at low temperatures, and deposit gummy material that appears as stringy threads and slime.

Additional types of milk spoilage result from the red pigment deposited in dairy products by *S. marcescens*, the blue or green pigment of *Pseudomonas* species, and the gray rot caused by certain *Clostridium* species. In gray rot, hydrogen sulfide (H_2S) from cysteine imparts a rotten-egg smell to milk, and the H_2S reacts with minerals to yield a gray or black sulfide compound. Spoilage due to wild yeasts is usually characterized by a pink, orange, or yellow coloration in the milk.

Milk is normally sterile in the udder of the cow, but contamination can occur as it enters the ducts leading from the udder, as well as from other unlikely sources. The **colostrum**, or first milk, is laden with organisms. Species of soilborne *Lactobacillus* and *Streptococcus* are acquired during the passage, together with various coliform bacteria from dust, manure, and polluted water. The milk may gain other organisms from dairy plant equipment and unsanitary handling of dairy products by plant employees (FIGURE 25.5).

FIGURE 25.5 **A Dairy Milking Facility.** A man is cleaning a dairy cow milking facility to prevent contamination of the milking equipment. © Inga Spence/Visuals Unlimited.
»» How would one sanitize a building such as a milking facility?

CONCEPT AND REASONING CHECKS 1

- **a.** Identify potential microbial sources leading to food spoilage.
- **b.** List the conditions that make foods susceptible to growth and spoilage by microorganisms.
- **c.** What type of microbial products lead to spoilage?
- **d.** Summarize how meats and seafood can become contaminated.
- **e.** Why are eggs and poultry so susceptible to spoilage?
- **f.** What types of organisms may grow in bakery products if they are not refrigerated?
- **g.** What type of microbial products lead to spoilage in bakery products and grains?
- **h.** Summarize the ways milk can become spoiled.

Colostrum: A fluid rich in antibodies and minerals that a mother animal produces after giving birth and before the production of true milk.

Chapter Challenge A

After returning to your apartment after being away for a week on Spring Break, you smell a funny odor and realize you left a carton of milk on the kitchen counter. Inspecting the milk, you notice it is lumpy and quite smelly. It has certainly spoiled!

QUESTION A:
What most likely caused the milk to spoil? What is the chemical reaction behind this process?

Answers can be found on the Student Companion Website in **Appendix D.**

KEY CONCEPT 25.2 Food Preservation Inhibits Foodborne Pathogens and Spoilage Microorganisms

Centuries ago, humans battled the elements to keep a steady supply of food at hand. Sometimes there was a short growing season; at other times, locusts descended on their crops; at still other times, they underestimated their needs and had to cope with scarcity. However, experience taught humans they could overcome these difficult times by preserving foods. Among the earliest methods was drying vegetables and strips of meat and fish in the sun. Foods also could be preserved by salting, smoking, and fermenting. Individuals could now trek far from their native habitat. Taking preserved food with them, they took to the sea and moved overland to explore new lands.

The next great advance did not come until the mid-1700s. In 1767, Lazaro Spallanzani attempted to disprove spontaneous generation by showing that beef broth would remain unspoiled after being subjected to heat. Nicholas Appert applied this principle to a variety of foods (MicroFocus 25.3).

MICROFOCUS 25.3: History

To Feed an Army

Part of Napoleon's genius was understanding the finer points of warfare, including how to feed an army. Recognizing that thousands of men-at-arms were a glut on the countryside, he broke up his army into smaller units that foraged on their own as they moved. When the time for battle neared, he reassembled his forces and engaged the enemy.

The shortcomings of this system became painfully clear when Napoleon crossed into northern Italy in 1800 and engaged the Austrians at the Battle of Marengo. Knowing the French army was scattered about, the Austrians attacked before Napoleon could bring his forces together. Disaster was averted only when units arrived on the flanks to repel the Austrians. Napoleon had learned an important lesson: The next time he went to war, food would go with him.

Included in Napoleon's plan to resurrect France was a ministry that encouraged industry by offering prizes for imaginative inventions. A winemaker named Nicholas Appert attracted attention with his process of preserving food. Appert placed fruits, vegetables, soups, and stews in thick bottles, then boiled the bottles for several hours. He used wax and cork to seal the bottles and wine cages to prevent inadvertent opening of the bottle. By 1805, Appert had set up a bottling industry outside Paris and had a thriving business.

The Ministry of Industry encouraged Appert to publish his methods and submit samples of bottled foods for government testing. The French navy took numerous bottles on long voyages and reported excellent food preservation. In 1810, Appert was awarded 12,000 francs for his invention. Two years later, Napoleon assembled hundreds of cannons, thousands of men, and countless bottles of food, as he marched off to war with Russia. That campaign would be a disaster due to another microbe, as described in another chapter.

Neither Appert nor any of his contemporaries were quite sure why food was being preserved, but it was clear that the spoilage could be retarded by prolonged heating. The significance of microorganisms as agents of spoilage awaited Pasteur's classic experiments with wine several generations later.

Through the centuries, preservation methods have had a common objective: to reduce the microbial population and maintain it at a low level until food can be consumed. Modern preservation methods are mere extensions of these principles. Though today's methods are sophisticated and technologically dynamic, advances in preservation processes are counterbalanced by the great volumes of food that must be preserved and the complexity of food products. Thus, the problems that early humans faced do not differ fundamentally from those confronting modern food technologists. In this section, we examine the methods of food preservation currently in use.

FIGURE 25.6 **The Industrial Canning Process.** An inspector watches over the canning process to ensure sanitary conditions are maintained. © Brian Klimek, The Laurinburg Exchange/AP Photos. »» How would tin cans be sterilized?

Heat Denatures Proteins

In a moist heat environment, proteins are denatured and lose their three-dimensional structure. As structural proteins and enzymes undergo this change, the organisms die. The most useful application of heat is in the process of canning. Shortly after Appert established the use of heat in preservation, an English engineer named Bryan Donkin substituted iron cans coated with tin for Appert's bottles. Soon he was supplying canned meat to the British navy. In the United States, the tin can was virtually ignored until the Civil War period. In the years thereafter, mass production began and soon the tin can became the symbol of prepackaged convenience.

Modern canning processes are complex. For produce, machines wash, sort, and grade the food product, and then subject it to steam heat for 3 to 5 minutes. This last process, called **blanching**, destroys many enzymes in the food product and prevents any further cellular metabolism from taking place. Then the food is peeled and cored, and its diseased sections are removed. Canning comes next, after which the air is evacuated and placed in a pressured steam sterilizer similar to an autoclave at a temperature of 121°C or lower, depending on the pH, density, and heat penetration rate (FIGURE 25.6).

The sterilizing process is designed to eliminate the most resistant bacterial spores, especially those of the genera *Bacillus* and *Clostridium*. However, this process is considered **commercial sterilization**, which is not as rigorous as true sterilization, and some spores may survive. Moreover, should a machine error lead to improper heating temperatures, a small hole allows airborne bacterial cells to enter, and a proper seal will not form, allowing contamination.

Contamination of canned food is commonly due to facultative or anaerobic bacterial species that produce gas and cause the ends of the can to bulge. Food microbiologists call a can a flipper if the bulge can be flattened easily. It is a springer if pushing the bulge pushes out the opposite end of the can. A soft swell occurs when both ends bulge. If neither end can be pushed in because of the large amount of gas, a hard swell is present. The organisms often responsible for gas production are *Clostridium* species, as well as coliform bacteria. Contamination is usually obvious since the spoiled product generally has a putrid odor.

Coliform bacteria: A group of gram-negative non-spore-forming rods that ferment lactose to acid and gas.

Growth of acid-producing bacterial species presents a different problem because spoilage cannot be discerned from the can's shape. Food has a flat-sour taste from the acid and has probably been contaminated by a *Bacillus* species, a coliform, or another acid-producing species that survived the heating.

The process of pasteurization was developed by Louis Pasteur in the 1850s to eliminate bacterial contamination of wines. His method was first applied to milk in Denmark about 1870, and by 1895 the process was widely employed. Although the primary object of **pasteurization** is to eliminate pathogenic bacterial species from milk, the process also lowers the total number of bacterial organisms, and thereby reduces the chance of spoilage. The more traditional method involves heating the milk in a large bulk tank at 63°C for 30 minutes. This is the **holding** (or **batch**) **method**, also known as the LTLT method for "low temperature, long time." Machines stir the milk constantly during the pasteurization to ensure uniform heating, and cool it quickly when the heating is completed. More concentrated products, such as cream, often are heated at the higher temperature of 69.5°C to ensure successful pasteurization. An innovative method of egg pasteurization has been recently introduced, as MicroFocus 25.4 explains.

The more modern method of pasteurization is called the **flash pasteurization method**. In this process, raw milk is first warmed using the heat of previously pasteurized milk. Machines then pass the milk through a hot cylinder at 71.6°C for a period of 15 seconds. Next, the milk is cooled rapidly, in part by transferring its heat to the incoming milk. This is the HTST method, for "high temperature, short time." It is useful for high-quality raw milk, in which the bacterial count is consistently low. Following pasteurization, a set of laboratory tests is performed to assay the quality of the milk and the success of pasteurization (TABLE 25.1).

MICROFOCUS 25.4: Biotechnology

"Bring on the Caesar Salad!"

Some people have it Hawaiian-style with pineapple and ham; some have it Italian-style with pasta and tomatoes; and some enjoy it Southwestern-style with roasted chili peppers. Some toss it with grilled chicken, flank steak, grilled calamari, or Cajun scallops. Regardless of the addition, however, the salad remains the same—Caesar salad.

© Jones & Bartlett Learning. Photographed by Kimberly Potvin.

For history buffs, the first Caesar salad is reported to have evolved on July 4, 1924, in the mind of Caesar Cardini, the proprietor of a restaurant (Caesar's Place) in Tijuana, Mexico. Cardini was desperate for a fill-in dish during a particularly busy day, so he threw together some Romaine lettuce, Parmesan cheese, lemon, garlic, oil, and raw eggs. His customers were enchanted.

Over the years, the reputations of both the salad and its inventor grew. There was only one problem, however—the eggs. For true Caesar salad, raw eggs are used to add creaminess to the dressing. But that became a problem when increasing cases of *Salmonella* infections were traced to raw eggs. Eggless Caesar dressings (with heavy cream) were tried, but it just wasn't the same.

Caesar salad aficionados, take heart—the pasteurized egg is on the way. Purdue University microbiologists have found that *Salmonella* can be eliminated by heating eggs in hot water or a microwave oven, then maintaining them at 57°C in a hot-air oven for one hour. Today, pasteurized eggs are readily available and the U.S. Department of Agriculture (USDA) says "in-shell pasteurized eggs may be used safely without cooking." They may be sold as liquid egg products or pasteurized in the shell. Indeed, the process works so well that the USDA has issued a new stamp (a red P surrounded by a red circle) certifying that such eggs meet standards for egg pasteurization established by the Food and Drug Administration (FDA).

Agricultural officials estimate that almost 75 billion eggs are produced for American consumption each year, and that over 2 million are infected with *Salmonella*. However, as pasteurized eggs become standard practice (as milk pasteurization already has) in food preparation, it will be safe to sample the cookie dough, or to have eggs over easy, or to enjoy Caesar salad the way it was meant to be enjoyed.

TABLE

25.1 Laboratory Tests Used to Assay the Quality of Milk

Test	Purpose	Importance
Phosphatase test	Determines whether phosphatase is present. Phosphatase is an enzyme normally destroyed during pasteurization.	To determine whether sufficient heat was used during pasteurization. If phosphatase is present, pathogens also might be present.
Reductase test	Estimates the number of bacterial cells in milk. The rate at which methylene blue is reduced to its colorless form is proportional to the number of cells present in a milk sample.	High-quality milk contains so few bacterial cells that a standard concentration of methylene blue will not be reduced in 6 hours. Low-quality milk has many bacterial cells, and methylene blue is reduced in 2 hours or less.
Standard plate count	Determines the total number of viable bacterial cells per milliliter of milk. Diluted milk is mixed with nutrient agar and incubated 48 hours; colonies are counted and the number of cells in the original sample is calculated.	The number per milliliter may not exceed 100,000 in raw milk before being pooled with other milk. It may not exceed 20,000 after pasteurization.
Test for coliforms	Determines the number of viable coliform bacteria per milliliter of milk. Similar to the standard plate count except that special media for coliform bacteria are used.	A positive coliform test indicates contamination with fecal material. The number per milliliter may not exceed established standards.
Test for pathogens	Detects the presence of pathogens. Methods depend on the pathogens suspected.	Helps locate the source of infectious agents that may be present in milk.

Sealed in sterile containers, the unopened milk will last about a month when refrigerated.

A newer method called **ultra high temperature (UHT) method** is used in some dairy plants. In this process, milk and milk products are subjected to a temperature of 140°C for 3 seconds. This ultra-high temperature results in milk (e.g., Parmalat) with an indefinite shelf life (storage normally is for about 6 months) as long as the container remains sealed. However, after opening, it must be kept in the refrigerator. These milk products have been popular in Europe, as they save on refrigeration costs. The product has a bit of a cooked taste, though.

The bacterial cells surviving pasteurization may be involved in spoilage. *Streptococcus lactis*, for example, grows slowly in refrigerated milk, and when its numbers reach 20 million per milliliter, enough lactic acid has been produced to make the milk sour. Organisms surviving the heat of pasteurization are described as **thermoduric**. Pasteurization has no effect on spores.

Low Temperatures Slow Microbial Growth

By lowering the environmental temperature, one can reduce the rate of enzyme activity in a microorganism, and thus, lower the rate of growth and reproduction. This principle underlies the process of refrigeration and freezing. Although the organisms are not killed, their numbers are kept low and spoilage is minimized. Ironically, the food is preserved by preserving the microorganisms.

Long before the invention of the refrigerator, the Greeks and Romans had partially solved the problem of keeping things cold. They simply dug a snow cellar in the basements of their homes, lined the cellar with logs, insulated it with heavy layers of straw, and packed it densely with snow delivered from far-off mountaintops. The compressed snow turned to a block of ice, and foods would remain unspoiled for long periods when left in this makeshift refrigerator.

The modern refrigerator at 5°C provides a suitable environment for preserving food without destroying appearance, taste, or cellular integrity (FIGURE 25.7). However, psychrotrophic microorganisms survive and cause green meat surfaces, rotten eggs, moldy fruits, and sour milk. Pathogens such as *Listeria monocytogenes* and *Yersinia enterocolitica* also grow at low temperatures.

When food is placed in the freezer at −10°C, ice crystals form rapidly. These crystals tear and

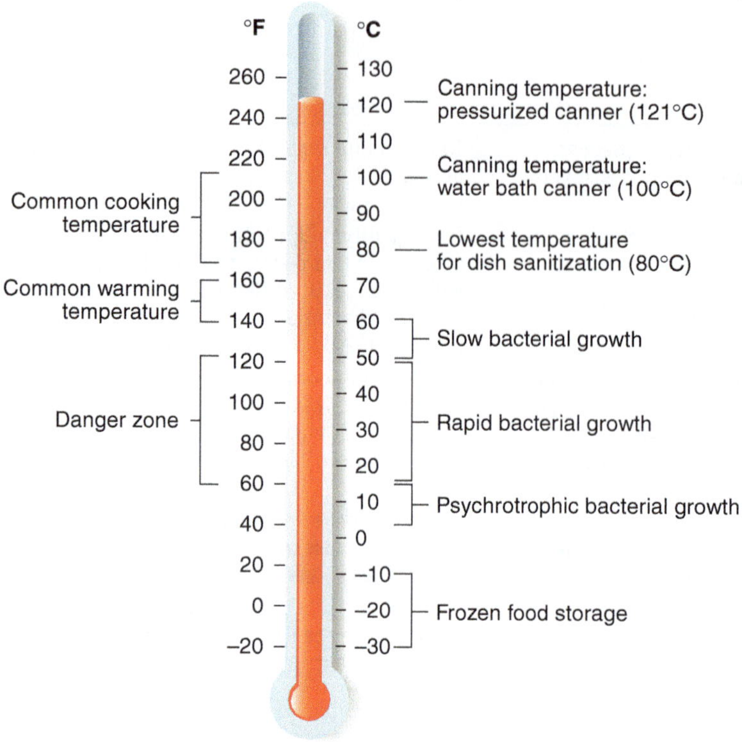

FIGURE 25.7 Important Temperature Considerations in Food Microbiology. Shown are several temperatures at which canning or sanitization procedures are carried out. »» Why is 15° to about 50°C called the "danger zone"?

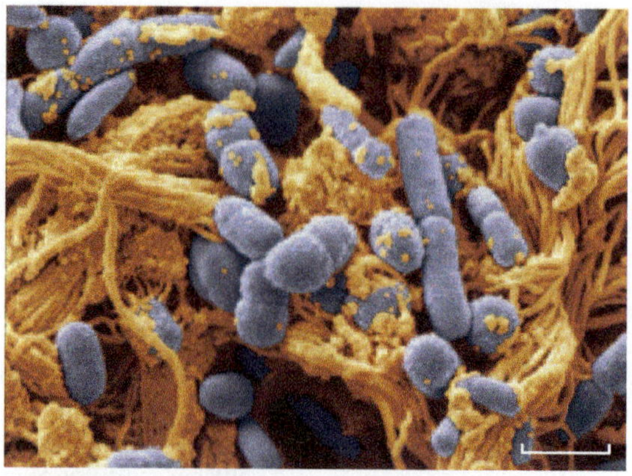

FIGURE 25.8 Food Contamination. False-color scanning electron micrograph of chicken skin from a retail chicken package. The skin is contaminated with *Salmonella* (purple cells) after having been kept for 24 hours at 22°C, showing how fast food can spoil outside the refrigerator. (Bar = 4 μm.) © Scimat/Photo Researchers, Inc. »» How did these spoilage bacteria come to contaminate the chicken skin?

shred microorganisms, killing a significant number of them in the process. However, many microorganisms survive, and the ice crystals that result are equally destructive to food cells. Therefore, when the food thaws, the surviving bacterial cells multiply quickly. Organisms such as staphylococci produce substantial amounts of enterotoxins, and *Salmonella* serotypes, streptococci, and other bacterial species grow to large numbers (FIGURE 25.8). Rapid thawing and cooking are recommended. Moreover, food should not be refrozen, because during thawing and refreezing, bacteria deposit sufficient enterotoxins to cause food poisoning the next time the food is thawed. Microwave cooking, which requires minimal thawing, may eliminate some of these problems.

Deep freezing at −60°C results in smaller ice crystals, and although the physical damage to microorganisms is less severe, their biochemical activity is reduced considerably. Small ice crystals do not damage food cells as severely as do the larger crystals formed at higher temperatures. Some food producers blanch their product before deep freezing, a process that further reduces the number of microorganisms.

Freezing has been a mainstay of preservation since Clarence Birdseye first offered frozen foods for retail purchase in the 1920s. Approximately 33% of all preserved food in the United States is frozen.

Drying and Osmotic Pressure Help Preserve Foods

The advantage of drying foods is best expressed by the phrase, "Where there is no water, there is no natural life." Indeed, dry foods cannot support microbial growth. In past centuries, people used the sun for drying, but modern technology has developed specific machinery for drying foods.

The spray dryer expels a fine mist of liquid, such as coffee, into a barrel cylinder containing hot air. The water evaporates quickly, and the coffee powder falls to the bottom of the cylinder. Another method uses a heated drum over which liquids such as soup are poured. The water evaporates rapidly, leaving dried soup to be scraped off the drum. A third machine uses a belt heater that exposes liquids such as milk to a stream of hot air. The air evaporates any water and leaves dried milk solids. Unfortunately, spore-forming and capsule-producing bacterial species can be problems because they resist drying.

MICROFOCUS 25.5: History

It Started with "Stomped Potatoes"

Freeze-drying, or lyophilization, involves the removal (sublimation) of water from a substance without going through a liquid state. Such freeze-dried foods last longer than other preserved foods and are very light, which makes them perfect for everyone from backpackers to astronauts. However, the origins of freeze-dried foods go back to the ancient Incas of Peru.

The basic process of freeze-drying food was known to the pre-Columbian Incas who lived in the high altitudes of the Andes Mountains of South America. The Incas stored some of their food crops, including potatoes, on the mountain heights. First, most of the moisture was crushed out of the potatoes when the Incas stomped on them with their feet. Then, the cold mountain temperatures froze the stomped potatoes and the remaining water inside slowly vaporized under the low air pressure of the high altitudes in the Andes.

Today, a similar procedure is still used in Peru for a dried potato product called chuno. It is a light powder that retains most of the nutritional properties of the original potatoes but, being freeze-dried, chuno can be stored for up to 4 years.

On the commercial market, the Nestlé Company was asked by Brazil in 1938 to help find a solution to save their coffee bean surpluses. Nestlé developed a process to convert the coffee surpluses into a freeze-dried powder that could be stored for long periods. The result was Nescafe® coffee, which was first introduced in Switzerland. However, during World War II the process of freeze-drying developed into an industrial process. Blood plasma and penicillin were needed for the armed forces and freeze-drying was found to be the best way to preserve and transport these materials.

In 1964, Nestlé developed an improved method of producing instant coffee by freeze-drying. Today, several other coffee producers use a similar process of preservation. In the late 1970s, freeze-drying was used for taxidermy, food preservation, museum conservation, and pharmaceutical production.

Today, more than 400 commercial food products are preserved by freeze-drying. In the medical field, freeze-drying is the method of choice for extending and preserving the shelf life of enzymes, antibodies, vaccines, pharmaceuticals, blood fractions, and diagnostics. The benefit of all lyophilized products is that they rehydrate easily and quickly because of the porous structure created by the freeze-drying process.

And to think—it all started with the Incas stomping on their potatoes.

During the past 20 years, freeze-drying, or **lyophilization**, has emerged as a valuable preservation method, although it actually has a long history (MicroFocus 25.5). In this process, food is deep frozen, and then a vacuum pump draws off the water in a machine like the one pictured in FIGURE 25.9. (Water passes from its solid phase [ice] to its gaseous phase [water vapor] without passing through its liquid phase [water].) The dry product is sealed in foil and easily reconstituted with water. Hikers and campers find considerable value in freeze-dried food because of its light weight and durability. Lyophilization is also a useful method for storing, transporting, and preserving bacterial cultures.

FIGURE 25.9 **An Industrial Lyophilizer.** This lyophilizer removes 500 pounds of product moisture in 24 hours of freeze-drying. Using vacuum and heat, water is drawn off from ice without passing through the liquid phase to produce the freeze-dried product. © Robert Longuehaye, NIBSC/SPL/Photo Researchers, Inc.

»» How are lyophilized products normally stored on the market shelf?

Osmotic pressure also helps preserve foods. When living cells are immersed in large quantities of a compound such as salt or sugar, water diffuses out of cells through the cell membranes and into the surrounding environment, where it dilutes the high concentration of the compound.

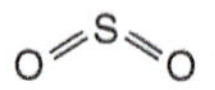
Sulfur dioxide

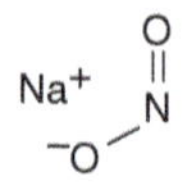
Sodium nitrite

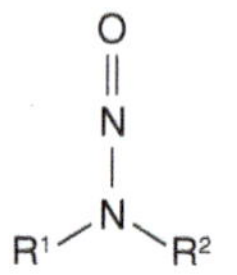
Nitrosamine

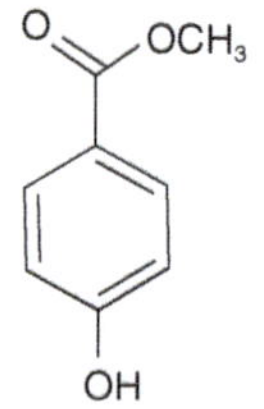
Methyl paraben

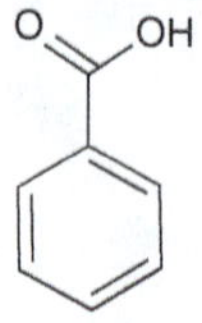
Benzoic acid

CH_3CH_2COOH
Propionic acid

This diffusion of water is called **osmosis**, and the tendency of water to move across a membrane is termed **osmotic pressure**.

Osmotic pressure can be used to preserve foods because water flows out of microorganisms as well as food cells. For example, in highly salted or sugared foods, microorganisms dehydrate, shrink, and die. Jams, jellies, fruits, maple syrups, honey, and similar products typify foods preserved by high sugar concentrations. Salted foods include ham, cod, bacon, and beef, as well as certain vegetables such as sauerkraut, which have the added benefit of containing large quantities of acid. It should be noted, however, that staphylococci tolerate salt and may survive to cause staphylococcal food poisoning.

Chemical Preservatives Inhibit Microbial Growth

For a chemical preservative to be useful in foods, it must be inhibitory to microorganisms while having the ability to be easily broken down and eliminated by the human body without side effects. These requirements have limited the number of available antimicrobial chemicals to a select few.

Organic Acids. A major group of chemical preservatives are **organic acids**, which are short-chain chemicals that are found in nature. However, most are now chemically synthesized for the food industry. On entering a microbial cell, the acid dissociates and releases protons (H^+) into the cytoplasm. To prevent the acidification of the cytoplasm, proton pumps must expel the protons out of the cell. This is an energy (ATP) exhausting process that slows or inhibits microbial growth. Due to their mode of action, organic acids are most useful in acidic foods with a pH lower than 5.5.

Benzoic acid was the first chemical to be approved by the FDA. Naturally found in cranberries, benzoic acid is used today to protect beverages, as well as jams and jellies from mold growth. **Propionic acid** occurs naturally in apples, strawberries, and grains. The acid is incorporated in wrappings for butter and cheese, and is added to breads and bakery products, where it inhibits the ropiness commonly due to *Bacillus* species and prevents the growth of fungi. Other organic acids in food add flavor while serving as preservatives. Examples are lactic acid in sauerkraut and yogurt, and acetic acid in vinegar.

Sulfites. The most common sulfite added to foods is **sulfur dioxide** (SO_2) which is used as a preservative for dried fruits, molasses, and juice concentrates. Again, dissociation of the acid form is important for preservation. Used in either gas or liquid form, the chemical retards color changes on the fruit surface and adds to the aesthetic quality of the product.

Nitrites. The curing of meats refers to food preservation and flavoring processes that add solutions of salt, sugar, spices, and nitrite. The addition of nitrites keeps meat looking pink, adds flavor and texture, and serves as antioxidants. They work best at lower pH and in the absence of air. Sodium nitrite inhibits a variety of potential bacterial pathogens, including *Clostridium botulinum* and *Listeria monocytogenes*. However, the reaction of nitrites with meat proteins at high temperature can produce nitrosamines that represent chemical carcinogens.

Other Antimicrobial Chemicals. Several other chemicals can be used as preservatives. This includes the hydroxybenzoic acids (commonly called parabens) that are active against molds and yeasts. Parabens can be found in such products as baked goods, beverages, jams and jellies, and salad dressings.

Simple phenolic compounds are sometimes used as preservatives. The process of smoking meats, fish, and poultry with hickory or other woods adds flavor and dries the food. The byproducts of smoke includes simple phenolic compounds, act as a preservative to inhibit a variety of potential bacterial pathogens.

Parabens and simple phenols affect membranes, perhaps by disrupting membrane integrity.

Foods May Be Irradiated for Pest and Pathogen Elimination

Though much of the public is apprehensive about foods exposed to radiation, various forms of radiation can be used to safely irradiate and sterilize foods. Taste and appearance are preserved by freezing food in liquid nitrogen and exhausting oxygen from the package before irradiation. Meat storage facilities use **ultraviolet (UV) light** to reduce surface contamination, and water can be treated with UV light when chlorine is not useful.

Gamma rays are used to extend the shelf life of fruits, vegetables, seafood, and poultry from several days to several weeks. This form of radiation also increases the distances fresh food can be transported and significantly extends the storage time of food in the home. Gamma rays are high-frequency forms of electromagnetic energy emitted by a radioactive isotope called cobalt-60. The emissions are not radioactive, and they cannot cause food to become radioactive. They kill microorganisms by reacting with and destroying microbial DNA.

Interest in gamma radiation for food preservation grew during the 1950s under President Eisenhower's Atoms for Peace program. For the next quarter-century, the FDA conducted extensive tests to determine whether the process was safe. TABLE 25.2 lists the irradiated foods that have been approved or pending approval. Realize that just because these foods have been approved for irradiation does not mean they are commercially available. Availability depends on the food and the company producing the food product. If the food has been irradiated it will have the radura symbol and "treated with irradiation" (or a similar label) on the packaged product (FIGURE 25.10).

TABLE

25.2 Foods Approved for Irradiation[a,b]

Year	Food
1964	Wheat flour and potatoes
1983	Dried spices
1985	Additional dried spices, herb and vegetable seasonings, dry and hydrated enzymes, pork
1986	Fruit and vegetables
1990	Poultry
1997	Beef, lamb, pork, and byproducts
2000	Fresh shell eggs, sprouting seeds, fruit and vegetable juices
2002	Imported fruit and vegetables (for pest treatment)
2005	Ready-to-eat foods (deli meats, frozen foods)
2008	Packaged salads (for pathogen treatment), molluscan shellfish (oysters, clams, mussels)
Pending	Crustacean shellfish (shrimp, crabs, lobster), unrefrigerated beef, lamb, pork, horsemeat, and byproducts

[a]Approvals by the U.S. Food and Drug Administration and Department of Agriculture.
[b]Data from: Food and Water Watch, Washington, DC.

Foodborne Disease Can Result from an Infection or Intoxication

Food may be a mechanical vector for infectious microorganisms or a culture medium for growth. People then are affected by the organism or by the toxin it has produced in the food. In the former case, a food infection is established; in the latter, a food poisoning or intoxication occurs.

Food infections are typified by typhoid fever, salmonellosis, cholera, and shigellosis, all of which are of bacterial origin. Amoebiasis and giardiasis represent protozoal foodborne diseases, while viral infections are exemplified by hepatitis A. Food **intoxications** include botulism, staphylococcal food poisoning, and clostridial food poisoning. Because full discussions of these diseases are presented in another chapter, we shall not examine them individually here.

In the United States, public health microbiologists estimate that between 2 and 10 million people are affected by foodborne illnesses annually. However, the number of outbreaks is increasing as globalization affects food production and transport (MicroFocus 25.6). Many illnesses require medical attention, but the vast majority of patients recover rapidly without serious complications. In many cases, the incident might have been avoided by taking some basic precautions. For example, unrefrigerated foods are a prime source of staphylococci and *Salmonella* serotypes, and perishable groceries such as meats and dairy products should not be allowed to warm up while other errands are performed. Also, a thermometer

FIGURE 25.10 **The Radura Symbol.** Food products that have been irradiated will have the radura label or some similar writing to indicate the food was irradiated. If a product only contains an irradiated ingredient, such as spices, no label is required. © Lack-O'Keen/ShutterStock, Inc.
»» Can you figure out the symbolism in the radura label?

MICROFOCUS 25.6: Public Health

Free Market Economy and Foodborne Diseases

© Stephen Coburn/ShutterStock, Inc.

Today, you can go into just about any supermarket and find fresh fruit that seems out of season. You can find fruits such as peaches and grapes in the middle of winter. Thanks to advances in agriculture and transportation, there has been a globalization of the food market. Peaches from Chile or raspberries from Mexico can be heading to U.S. markets within hours after being picked. It's great to have fresh fruits and vegetables, but it brings along the risk of a foodborne illness.

About 70% of many fresh fruits and vegetables we find in our local supermarket come from other countries. Sometimes these countries do not have the regulated and safe food practices found in the United States and other developed countries. Although outbreaks of foodborne illnesses are rare, they are occurring with greater regularity every year. Actual data can be hard to collect because many people never visit a doctor for a foodborne illness.

The Centers for Diseases Control and Prevention (CDC) states that although foodborne infections have decreased by nearly 25% since 1996, "continued investments are essential to detect, investigate, and stop outbreaks promptly in order to protect our food supply."

Detecting foodborne outbreaks is not easy because fresh produce is distributed across the nation and not to just one local community. In fact, physicians and local health departments may not see a local outbreak as part of a national event. Most local health departments are accustomed to responding to a foodborne outbreak at a social function or commercial establishment. Globalization now means that local, state, and national food safety programs need to coordinate their surveillance methods and establish clear communication channels to properly recognize and quickly respond in a timely fashion to a foodborne outbreak.

Importantly, should you shy away from imported produce? Certainly not. In most all cases, proper washing of fresh fruits and vegetables, or cooking produce, will eliminate almost any chance of contracting a foodborne illness. Keep consuming those fruits and vegetables—five times a day!

should be used to ensure that the refrigerator temperature is at or below 5°C at all times.

Another way to avoid foodborne disease is to cover any skin boils while working with foods because boils are a common source of staphylococci. The hands should always be washed thoroughly before and after handling raw vegetables or salad fixings to avoid cross-contamination of other foods. It is wise to cook meat from a frozen or partly frozen state; if this is impossible, the meat should be thawed in the refrigerator. Cutting boards should be cleaned with hot, soapy water after use, and old cutting boards with cracks and pits should be discarded. MICROINQUIRY 25 quizzes you on your sanitary precautions.

Studies indicate that leftovers are implicated in most outbreaks of foodborne disease. It is therefore important to refrigerate leftovers promptly and keep them no more than a few days. Thorough reheating of leftovers, preferably to boiling, also reduces the possibility of illness.

Many instances of foodborne disease occur during the summer months, when foods are taken on picnics where they cannot be refrigerated. As a general principle, dairy foods, such as custards, cream pies, pastries, and deli salads should be excluded from the picnic menu. For outdoor barbecues, one dish should be used for carrying hamburgers to the grill and another dish for serving them. Many of these principles apply equally well to fall and winter tailgate parties.

Over 90% of botulism outbreaks reported to the CDC are traced to home-canned food. To prevent this sometimes fatal foodborne disease, health officials urge homemakers to use the pressure method to can foods. Reliable canning instructions

MICROINQUIRY 25

Keeping Microorganisms Under Control

How knowledgeable are you concerning food safety in your home or apartment? If you are, do you actually practice what you know? Take the following quiz (honestly) and then we will see where microbes lurk, and where you need to eliminate or keep them under control. Answers can be found on the Student Companion Website in **Appendix D.**

25.1. Your refrigerator should be kept at what temperature?
a. 0°C
b. 5°C
c. 10°C
d. Do not know

25.2. Frozen fish, meat, and poultry products should be defrosted by _____.
a. setting them on the counter for several hours
b. microwaving
c. placing them in the refrigerator

25.3. After cutting raw fish, meat, or poultry on a cutting board, you can safely _____.
a. reuse the board as is
b. wipe the board with a damp cloth and reuse it
c. wash the board with soapy hot water, sanitize it with a mild bleach, and reuse it

25.4. After handling raw fish, meat, or poultry, how do you clean your hands?
a. Wipe them on a towel.
b. Rinse them under hot, cold, or warm tap water.
c. Wash them with soap and warm water.

25.5. When was the last time you sanitized your kitchen sink drain and garbage disposal?
a. Last night
b. Several weeks ago
c. Several months ago
d. Cannot remember

25.6. Leftover cooked food should be _____.
a. cooled to room temperature before being put in the refrigerator
b. put in the refrigerator immediately after the food is served
c. left at room temperature overnight or longer

25.7. How do you clean your kitchen counter surfaces that are exposed to raw foods?
a. Use hot water and soap, then a bleach solution.
b. Use hot water and soap.
c. Use warm water.

25.8. How was the last hamburger you ate cooked?
a. Rare
b. Medium
c. Well-done

25.9. How often do you clean or replace your kitchen sponge or dishcloth?
a. Daily
b. Every few weeks
c. Every few months

25.10 Normally dishes in your home are cleaned _____.
a. by an automatic dishwasher and air-dried
b. after several hours soaking and then washed with soap in the same water
c. right away with hot water and soap in the sink and then air-dried

So, how did you do?

should be obtained and followed stringently. Foods suspected of contamination should not be tasted to confirm the suspicion, but should be discarded immediately. When in doubt, boiling the food for a minimum of 10 minutes and thoroughly washing the utensil used to stir the food are recommended. Bulging or leaking cans must be discarded in a way that will not endanger other people or animals.

Milk is an unusually good vehicle for the transmission of pathogenic microorganisms because its fat content protects organisms from stomach acid, and, being a fluid, it remains in the stomach a relatively short period of time. The diseases of cows transmitted by milk include bovine tuberculosis, brucellosis, listeriosis, and Q fever. The biggest threat comes from drinking unpasteurized milk. In late 2002 and early 2003, the CDC received reports of a multistate outbreak of *Salmonella enterica* serotype Typhimurium that affected more than 94 individuals who had consumed raw, unpasteurized milk. Another case was reported in 2007, arising from the drinking of raw milk contaminated with *E. coli* O157:H7. Of some 140 people who drank the milk, 18 became ill and 4 developed hemolytic uremic syndrome.

Another milkborne organism of significance is *Campylobacter jejuni*, the cause of campylobacteriosis. In 2001, 75 persons in Wisconsin contracted campylobacteriosis from consuming raw, unpasteurized milk. Health officials estimate

Campylobacter jejuni is present in the intestinal tracts of about 40% of dairy cattle.

Despite the known association of raw milk with disease-causing organisms, some consumers believe that raw milk is of better quality than pasteurized milk. However, any health or nutritional benefits from consuming raw milk have not been proved scientifically. Public health microbiologists seek to limit milkborne disease by inspecting food and dairy plants regularly, and then making recommendations on improved sanitary practices.

HACCP Systems Attempt to Identify Potential Contamination Points

Fueled by consumer awareness, the entire food industry has been placed under a food-safety spotlight overseen by the FDA and the USDA.

Among the most important food safety systems is **Hazard Analysis and Critical Control Point (HACCP)**, a set of scientifically based safety regulations enforced in the seafood, meat, and poultry industries. In an HACCP system, manufacturers identify individual processing points, called **critical control points (CCPs)** where the safety of a product could be affected, such as by contaminating microorganisms. The CCPs are supervised to ensure that any hazards associated with the operation are contained or, preferably, eliminated (FIGURE 25.11). When all possible hazards are controlled at the CCPs, the safety of the product can be assumed without further testing or inspection.

The standard regulations require food processors to monitor and control eight key sanitation areas, including the condition and cleanliness of utensils, gloves, outer garments, and other food contact surfaces; the prevention of cross-contamination from raw products and unsanitary objects to foods; and the control of employee health conditions that could result in food contamination.

CONCEPT AND REASONING CHECKS 2

- **a.** Distinguish between the various forms of heat in the process of pasteurization.
- **b.** How does refrigeration temperature slow down the growth of microbes?
- **c.** How do lyophilization and osmotic pressure preserve foods?
- **d.** How do chemical preservatives work to prevent spoilage?
- **e.** What type of radiation is used to irradiate foods?
- **f.** How do food infections differ from food intoxications?
- **g.** What are critical control points in the seafood, meat, and poultry industries?

FIGURE 25.11 **Meat Inspections.** A meat inspector checks beef prior to sale. The long-standing "sniff-and-poke" method of inspection is being replaced by a newly instituted HACCP system of ensuring meat safety. Courtesy of USDA. »» What CCP is being supervised in this photo?

Chapter Challenge B

At the dairy or factory site, samples from the bulk milk are tested for antibiotics and temperature before the milk goes to the processing area. Farm milk samples are then analyzed for their milkfat and protein content and a bacterial cell count is carried out. If milk does not meet quality standards it is rejected. Whole milk, once approved for use, is pumped into storage silos where it undergoes pasteurization before further processing.

QUESTION B:

What is pasteurization attempting to do? So, what happened then with the carton of spoiled milk you discovered on your return home?

Answers can be found on the Student Companion Website in **Appendix D.**

KEY CONCEPT 25.3 Many Foods Are the Product of Microbial Metabolism

Over the centuries, social customs and traditions have brought acceptance of a variety of foods produced by microorganisms. Some individuals regard these foods as "spoiled," but to many people, the food is deliciously "fermented" (FIGURE 25.12).

Many Foods Are Fermented Products

Fermentation precedes human history, as many fruits naturally ferment. The first known examples of humans controlling fermentation date back to at least 3000 BC, when the Babylonians and Egyptians fermented beverages like wine and beer. The strict biochemical definition of **fermentation** is the process for producing ATP using endogenous organic compounds as both electron donors and acceptors—exogenous electron acceptors are absent in the absence of oxygen gas. However, **food fermentations** may occur with or without oxygen gas. They are biochemical processes that flavor foods, alter their texture, and perhaps detoxify the product.

Fermentation can produce a number of end (waste) products, depending on the organism doing the fermentation. End products include lactic and acetic acids, ethanol (ethyl alcohol), or other reduced end products.

FIGURE 25.12 **An Array of Foods and Beverages Produced by Microorganisms.** Many foods and beverages are the result of microbial fermentations. © Jones & Bartlett Learning. Photographed by Kimberly Potvin.
»» What is common to all these fermentation processes?

Fermented foods have four characteristics:

- Foods are less vulnerable to extensive spoilage than unfermented foods because of the acids or alcohol produced.
- Foods are less likely to be vectors of foodborne illness because of the acids and/or alcohol present.
- Foods may become enriched with protein, amino acids, and vitamins.
- Foods are enriched in their taste, texture, and flavor.

Here, we will mention a few typical fermented food products.

Fermented Vegetables. Sauerkraut (German for "sour cabbage") is not only a well-preserved and tasty form of cabbage, but also nutritionally sound. For instance, the vitamin C content of sauerkraut is equivalent to that of citrus fruits, and sauerkraut often was taken on British sea voyages to prevent scurvy when citrus fruits were too expensive.

Scurvy: A vitamin C deficiency that causes spongy gums and bleeding into the skin.

Sauerkraut is prepared commercially by:

- Adding salt to shredded cabbage and packing the cabbage tightly
- Fermentation starts with species of *Klebsiella* and *Enterobacter* that produce more anaerobic conditions.
- This sets up the perfect environment for species of *Leuconostoc*, a gram-positive coccus found naturally in the cabbage. The bacterial cells ferment carbohydrates in the plant cells and produce acetic and lactic acids. As days go by, the acids lower the pH to about 3.5 and the sauerkraut would be considered "pasteurized."
- Species of *Lactobacillus* then proliferate, and the additional lactic acid they produce reduces the pH even more. The salt helps retard mold contamination while drawing juices out of the plant cells. A compound called diacetyl (the flavoring agent in butter) is produced by *Leuconostoc*, adding aroma and flavor.

In the United States, "pickle" is practically synonymous with "pickled cucumber." Over 37 types of dill, sour, and sweet pickles have been categorized, but essentially, all cucumber fermentations are similar. Cucumbers are placed in brine, a salt solution of 8% or higher, at which point the cucumber changes color from bright green to olive green.

Three groups of microorganisms are important to the fermentation and curing of cucumbers. *Enterobacter aerogenes,* a gram-negative rod, produces large amounts of CO_2, which takes up all the air space and establishes anaerobic conditions. *Lactobacillus* and *Leuconostoc* species then dominate and form abundant amounts of acid, which softens the tissues and sours the cucumbers. Finally, certain yeasts grow and establish many flavors associated with ripe pickles. Dill, garlic, and other herbs and spices are added to finish the product.

Most pickles are heat pasteurized or further acidified to increase their shelf life, with the exception of "kosher-style" pickles, which are given no further treatment. Pickled peppers, tomatoes, and other vegetables undergo a somewhat similar process.

Vinegar Fermentation. Vinegar (derived from the French *vinaigre* for "sour wine") is a fermented liquid traditionally made by the spontaneous souring of wine. Pasteur's ideas concerning the germ theory of disease were built in part from his examinations of the souring of French wines.

A widely used method for industrial vinegar production uses yeasts to ferment fruit juice to alcohol until the alcohol concentration is about 10% to 20%. Machines then spray the alcoholic juice into a tank containing the bacterial species *Acetobacter aceti* growing on the surface of wood shavings, gravel, or other substrate. In this second aerobic fermentation, as alcohol percolates through, bacterial enzymes convert it to acetaldehyde and then acetic acid. The vinegar

$$\underset{\text{Ethanol}}{CH_3CH_2OH} + O_2 \rightarrow \underset{\text{Acetic acid}}{CH_3COOH} + H_2O$$

recirculates several times before collection at the bottom of the tank. Residual alcohol evaporates in the heat, and the product usually has an acetic acid content of about 3% to 5%. The flavor of vinegar is determined by oils, sugars, and other compounds produced by the bacterial cells, plus the residue of organic compounds in the wine from which it was made.

Other Fermented Foods. Several other fermented foods also are worthy of note. One example, soy sauce, is made from roasted soybeans and wheat inoculated with the fungus *Aspergillus oryzae* and allowed to stand for 3 days. The fungus-covered product, called koji, then is added to a solution of salt and microorganisms, and aged for about a year (FIGURE 25.13). During this time, lactobacilli produce acid, and yeasts produce small amounts of alcohol. Together with the fungal products, the acid and alcohol determine the flavor. The liquid pressed from the mixture is soy sauce.

Another example is fermented sausages. These are produced generally as dry or semidry products and include pepperoni from Italy, thuringer from Germany, and polsa from Sweden. The process for producing these products begins with the addition of nitrites, curing agents, and spices to ground meats, followed by stuffing the meat into casings and incubating them at warm temperatures. Mixed acids produced from carbohydrates in the meat give the sausage its unique flavor and aroma.

Cocoa and coffee also owe their flavor, in part, to microorganisms. Cocoa derives some of its taste from the microbial fermentation that helps remove cocoa beans from the pulp covering them in the pod. Likewise, coffee is believed to obtain some of its flavor from the yeast and bacterial fermentation of coffee berries when the beans are soaked in tanks of water to loosen the berry skins before roasting.

FIGURE 25.13 **The Production of Soy Sauce.** Roasted soybeans are inoculated with *Aspergillus oryzae* and then mixed with a salt solution containing other microorganisms. © Pacific Press Service/Alamy Images. »» Why would an opened bottle of soy sauce that has been left unrefrigerated become slightly bitter tasting?

Many Milk Products Are the Result of Fermentation

Over the centuries dairy fermentations have been used to produce fermented milk products and cheeses.

Fermented Milk Products. Fermented milk products have assumed a key place in our diet. The sour milks are typical examples of fermented milk products. Buttermilk is made by adding starter cultures of *Streptococcus cremoris* and *Leuconostoc citrovorum* to vats of skim milk. The *Streptococcus* ferments lactose to lactic and acetic acids, and the *Leuconostoc* continues the fermentation to yield various aldehydes and ketones, and the compound diacetyl. These substances, especially diacetyl, give buttermilk its flavor, aroma, and acidity. For sour cream, the same process is used, except that pasteurized light cream is the starting point.

Acidophilus milk is produced in much the same way as buttermilk, except that the skim milk is inoculated with *Lactobacillus acidophilus*. This bacterium is a normal member of the human intestinal flora. Many healthcare practitioners believe that the lactobacilli in acidophilus milk augment naturally occurring lactobacilli and help the digestion process, while keeping molds in check. Another type of acidophilus milk, sweet acidophilus milk, lacks the sour taste. It is prepared by adding *L. acidophilus* to pasteurized milk.

Yogurt is a form of sour milk, made by adding dry milk solids to boiled milk to achieve a custard-like consistency. The two starter cultures are *Streptococcus thermophilus* and *Lactobacillus bulgaricus*. Many microbiologists hold that these bacterial species in yogurt support good health similar to the bacterial species in acidophilus milk.

Cheese Manufacture. Cheese production begins when the casein curdles out of milk. Usually this accompanies a souring of the milk by streptococci, but the process may be accelerated by adding rennin, an enzyme obtained from the stomach lining of a calf (FIGURE 25.14). The milk curd is essentially an unripened cheese. It may be marketed as cottage cheese, or pot cheese. Cream cheese also is unripened cheese with a butterfat content of up to 20%.

To prepare ripened cheese, the milk curds are washed, pressed, sometimes cooked, and then cut to the desired shape. Often the curds are salted to add flavor, control moisture, and prevent contamination by molds. If Swiss cheese is to be made, two bacterial species grow within the cheese: *Lactobacillus* species, which ferment the lactose to lactic acid; and *Propionibacterium* species, which produce organic compounds and carbon dioxide, which seeks out weak spots in the curd and accumulates as holes, or eyes. Cheddar cheese is scalded at a lower temperature than Swiss. Cheddar and Swiss are examples of cheese ripened internally by bacteria; Provolone, Edam, and Gouda are others.

FIGURE 25.14 **Cheese Manufacture.** The process of cheese manufacture begins by adding rennin to a vat of milk. © David R. Frazier Photolibrary, Inc./Alamy Images.
»» Why is rennin added in cheese manufacturing?

■ Aldehyde: An organic compound containing a carbon atom double bonded to an oxygen atom, and single bonded to a carbon atom and a hydrogen atom.

■ Ketone: An organic compound containing a carbon atom double bonded to an oxygen atom, and single bonded to two carbon atoms.

Another group of cheeses is somewhat softer in texture, a characteristic deriving from the partial breakdown of the protein curds by microbial enzymes. Growth takes place primarily at the surface of these cheeses, and the products tend to be pungent. Within the group of soft cheeses are Muenster, Port du Salut, and Limburger. Yeasts and species of the gram-positive rod *Brevibacterium* are among the microbes on the surface of the cheese. The rind of the cheese is derived from microbial pigments.

Milk from animals other than cows can be used to make cheeses. Goat cheese (chèvre; French for *goat*) uses the milk from domesticated goats. A bacterial starter culture of rennet and streptococci coagulate the milk (FIGURE 25.15). The water in the milk is strained away as whey and the remaining casein protein is ripened to obtain the structure and flavor of the cheese. The higher proportion of fatty acids in goat's milk imparts a tart flavor to the cheese.

The mold-ripened cheeses are represented by camembert. Camembert cheese is made by dipping salted curds into *Penicillium camemberti* spores to stimulate a surface growth. The fungus grows on

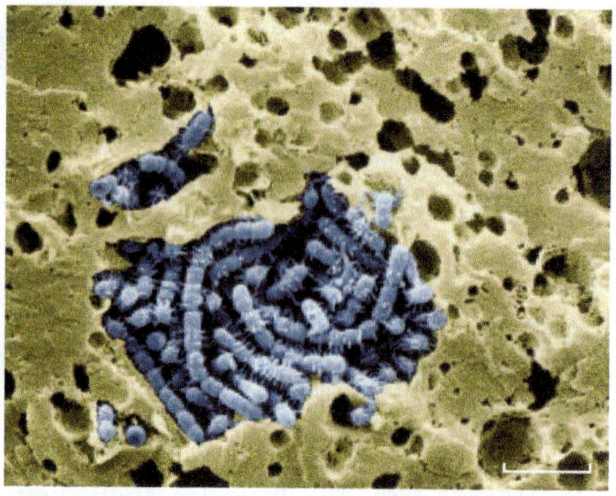

FIGURE 25.15 **Goat Cheese.** A false color scanning electron micrograph of goat cheese showing the lactic acid streptococci (blue). (Bar = 10 μm.) © Scimat/Photo Researchers, Inc. »» What do the streptococci do for the cheese?

the outside of the curd and digests the proteins, thus softening the curd. Stilton is another mold-ripened cheese where the mold penetrates cracks within the curd, creating its distinctive veins. Such cheeses have a unique community of microbes as the Investigating the Microbial World 25 reports.

CONCEPT AND REASONING CHECKS 3

a. Identify the common end products to fermented foods, such as sauerkraut, vinegar, and soy sauce.

b. What types of microbes can be found in cheeses?

Chapter Challenge C

In some cases, milk is purposely allowed to "sour." Among the raw or pasteurized milk products are: kefir [a fermented milk drink made with a mixture of bacteria and yeasts mixed with protein, sugars, and lipids]; yogurts; buttermilk; cottage cheese, feta, fresh mozzarella, mascarpone, cream cheese, and all sorts of other soft to hard cheeses.

QUESTION C:

Explain why spoiled milk is not fit for consumption but so many soured milk products are fine and enjoyed by millions of people every day.

Answers can be found on the Student Companion Website in **Appendix D.**

Investigating the Microbial World 25

The Microbial Ecology of Cheese

Globally, more than 20 million metric tons of cheese are produced annually, making cheese a major food product. Much of its magic comes from the wide-ranging variety of complex flavors, textures, and forms that different traditional cheeses possess especially when compared to the modern, controlled-inoculation cheeses. Often the traditional cheeses are the result of the dozens of different molds and/or bacterial species that have been added to the cheese during the production process. Modern microbiology has yet to fully explain the role of all microbes in cheese production.

- **OBSERVATIONS:** Stilton cheese is an internally mold-ripened semisoft blue cheese made from pasteurized cow's milk. The process involves seeding the milk with *Lactococcus lactis* and then the ripening is promoted by the mold *Penicillium roqueforti* as well as yeasts and other bacterial species. How the members of this complex microbiota interact in the ripening process remains to be discovered.
- **QUESTION:** ***What is the bacterial community structure in Stilton cheese?***
- **HYPOTHESIS:** The bacterial species composing the microbiota of Stilton cheese are randomly distributed on and throughout the cheese matrix. If so, then molecular characterizations to pinpoint the location of the microbes will indicate a jumbled community structure.
- **EXPERIMENTAL DESIGN:** Samples of Stilton cheese from 16 different areas of production in the United Kingdom were initially used. One sample was eventually chosen for analysis. Identification of most likely species and community structure were analyzed by:

Investigating the Microbial World 25 (continued)

- **EXPERIMENT 1:** PCR amplification and gel electrophoresis to identify most likely species.
- **EXPERIMENT 2:** Fluorescently labeled probes to detect and localize the distribution of specific microbial species (ribosomal DNA sequences) within different regions of the cheese matrix.
- **RESULTS:**

EXPERIMENT 1: The molecular analysis of samples extracted directly from the cheese corresponded to portions of 16S rDNA sequences from members of the domain Bacteria: *Lactococcus lactis, Enterococcus faecalis, Lactobacillus curvatus, Staphylococcus equorum, Leuconostoc mesenteroides,* and *Lactobacillus plantarum.*

EXPERIMENT 2: Fluorescently labeled probes were prepared from *Lactococcus lactis, Leuconostoc mesenteroides,* and *Lactobacillus plantarum. Lactococcus lactis* made up about 70% of the colonies in the core with *Leuconostoc mesenteroides* making up another 15%. Unidentified cocci made up the other 15%. Extensive colonies were found in the margins of the veins but none could be identified with the three probes developed. Being that they consisted of rods, they may be *Lactobacillus curvatus.* Nonpathogenic staphylococci were only on the vein surface while *Leuconostoc mesenteroides* colonies were found in the central portion of the veins. The blue *Penicillium roqueforti* was found in the veins. *Lactobacillus plantarum* and presumably *Lactobacillus curvatus* were present only under the crust.

- **CONCLUSIONS:** Based on the results, the hypothesis is not supported. There was a differential distribution of species found in the cheese matrix, suggesting there may be ecological reasons for the bacterial species being found were they were.

QUESTION 1: ***Would one expect to find all these bacterial species considering what species is used to initially seed the pasteurized milk culture and the ripening process? Where might these other species have come from?***

QUESTION 2: ***Using the image of Stilton cheese, label where the different bacterial species and P. roqueforti were located.***

QUESTION 3: ***Different batches of, and often poor quality, Stilton cheeses vary in their flavor and taste. Based on the results, account for these alterations in flavor and quality.***

Answers can be found on the Student Companion Website in **Appendix D.**

Adapted from: Ercolini, D., Hill, P. J., and Dodd, C. E. R. (2003). *Appl. Environ. Microbiol.* **69**(6): 3540–3548.

SUMMARY OF KEY CONCEPTS

25.1 Food Spoilage Is Generally a Result of Microbial Contamination and Growth

1. Contaminating microbes can enter food from the air, soil, shellfish, and animals, as well as through human handling of foods.
2. Certain conditions, such as water content, pH, physical structure, and chemical composition determine the extent of **food spoilage** because they influence the growth of microorganisms. (Fig. 25.2)
3. Products of spoilage include alcohol, foul-smelling products such as indole and skatole, and acids and gases. Capsules and pigments also are involved with spoilage.
4. In meats and seafood, spoilage organisms enter during processing; in seafood, water often is the source.
5. In poultry, the spoilage may reflect human contamination, but often it is due to members of the genus *Salmonella* that infected the bird.

6. Bakery ingredients generally bring contaminants to bread and other bakery products.
7. Grains, including peanuts and rye, may be spoiled by toxin-producing fungi.
8. Dairy products are spoiled by microorganisms surviving **pasteurization**, such as curd, capsule, and pigment producers.

25.2 Food Preservation Inhibits Foodborne Pathogens and Spoilage Microorganisms

9. To preserve food from spoilage, a number of methods are used, including heat. The most useful application of heat is in the process of canning, while pasteurization is used for milk and other liquid products.
10. Low temperatures are achieved in the refrigerator and freezer. Low temperatures do not kill all microorganisms contaminating a refrigerated or frozen food product.
11. Drying is useful because water is an absolute necessity for life, and dried foods, therefore, are unable to support microbial life. An ever-increasing list of foods preserved by freeze-drying is appearing on the market.
12. Foods high in sugar or salt preserve foods by drawing water out of any microorganisms that might be contaminating the food product. Various chemicals, such as **propionic acid** and **benzoic acid**, are used to preserve foods.
13. **Ultraviolet light** and **gamma rays** typify the radiation used in processing certain foods.
14. Individuals can contract a foodborne disease either through an infection or by ingesting a microbial toxin.
15. The **Hazard Analysis and Critical Control Point** (**HACCP**) is a set of scientifically based safety regulations that attempt to identify food processing points that are subject to contamination.

25.3 Many Foods Are the Product of Microbial Metabolism

16. Certain foods "spoiled" by microorganisms have come to be accepted as the norm. Among these microbial products are sauerkraut, vinegar, and fermented sausages. The spoilage in these foods causes no harm to consumers, and the foods reflect the helpful activities that microorganisms perform to add to the quality of our lives. Other foods involving fermentation products include soy sauce, sausages, cocoa, and coffee.
17. Buttermilk, sour cream, and yogurt are products where microbial action produces a fermented milk product. Numerous dairy products, such as cheeses, are also products of microorganism fermentation.

CHAPTER SELF-TEST

For **STEPS A–D**, answers to even-numbered questions and problems can be found in **Appendix C** on the Student Companion Website at **http://microbiology.jbpub.com/10e**. In addition, the site features eLearning, an online review area that provides quizzes and other tools to help you study for your class. You can also follow useful links for in-depth information, read more MicroFocus stories, or just find out the latest microbiology news.

STEP A: REVIEW OF FACTS AND TERMS

Multiple Choice

Read each question carefully, then select the ***one*** answer that best fits the question or statement.

1. Foods can become spoiled by _____.
   A. human handling
   B. rodent or insect contamination
   C. contamination during packaging
   D. All the above (**A–C**) are true.
2. Which one of the following would NOT be considered a highly perishable food?
   A. Milk
   B. Eggs
   C. Meat
   D. Dried rice and beans
3. All of the following are spoilage products produced by microorganisms *except* _____.
   A. lysozyme
   B. hydrogen sulfide
   C. ethyl alcohol
   D. skatole
4. Meats and fish normally become contaminated by _____.
   A. handling and processing
   B. the muscle tissue from the living animal
   C. airborne particles
   D. preserving agents
5. The bacterial genus _____ is a common source of poultry and egg contamination.
   A. *Escherichia*
   B. *Listeria*
   C. *Salmonella*
   D. *Clostridium*
6. The souring of bakery cream fillings and toppings is due to microbial _____.
   A. protein degradation
   B. acids
   C. carbohydrate hydrolysis
   D. fat breakdown
7. The fungal aflatoxins are typically found in _____.
   A. rye plants
   B. stored grains
   C. rotten eggs
   D. spoiled meat
8. Milk spoilage is usually due to species of _____.
   A. *Streptococcus*
   B. *Escherichia*
   C. *Salmonella*
   D. *Pseudomonas*
9. The flash method of pasteurization subjects milk to a temperature of ___°C for ___ seconds.
   A. 62.9; 30
   B. 71.6; 15
   C. 140; 3
   D. 161; 5

10. To keep foods fresh longer or frozen, a home refrigerator should keep foods at ___°C, while the freezer should be kept at ___°C.
    - **A.** 4; –10
    - **B.** 10; –4
    - **C.** 10; 0
    - **D.** 32; 4
11. Lyophilized foods have been _____.
    - **A.** heat-dried
    - **B.** irradiated
    - **C.** chemically treated
    - **D.** freeze-dried
12. Which one of the following is NOT a chemical preservative?
    - **A.** Propionic acid
    - **B.** Penicilloic acid
    - **C.** Sulfur dioxide
    - **D.** Benzoic acid
13. Commercial foods can be sterilized using _____.
    - **A.** gamma rays
    - **B.** microwaves
    - **C.** UV light
    - **D.** X rays
14. Food intoxications include all the following *except* _____.
    - **A.** staphylococcal food poisoning
    - **B.** botulism
    - **C.** hepatitis A
    - **D.** salmonellosis
15. Food safety in the United States is under the supervision of the _____.
    - **A.** CDC
    - **B.** USDA
    - **C.** DOE
    - **D.** HHS
16. All of the following foods or beverages involve a fermentation step *except* _____.
    - **A.** sauerkraut
    - **B.** vinegar
    - **C.** soft drinks
    - **D.** pickles
17. Fermented dairy products include _____.
    - **A.** eggs
    - **B.** buttermilk
    - **C.** yogurt
    - **D.** Both **B** and **C** are correct.

Identification

On completing this chapter on food microbiology, test your knowledge of its contents by using the following syllables to compose the term that answers the clue. Each term is a genus of microorganism important in food microbiology. The number of letters in the genus is indicated by the dashes, and the number of syllables in the genus is shown by the number in parentheses. Each syllable is used only once.
A A A AS AS BA BAC BAC CE CEPS CHLA CIL CLAV CLO CLO CO COC CUS DI DO EN GIL I I I LA LEU LUS LUS LUS MO MON MY NEL NOS O PER PSEU RA SAL SER STREP STRID STRID TER TER TER TI TO TO TOC US US

18. Common poultry contaminant (4) __ __ __ __ __ __ __ __ __ __
19. Destroyed in canning (4) __ __ __ __ __ __ __ __ __ __ __
20. Red pigment in bread (4) __ __ __ __ __ __ __ __
21. Causes psittacosis (4) __ __ __ __ __ __ __ __ __
22. Sours dairy products (4) __ __ __ __ __ __ __ __ __ __ __ __ __
23. Spore-forming contaminant (3) __ __ __ __ __ __ __ __
24. Used to make vinegar (5) __ __ __ __ __ __ __ __ __ __ __
25. Sauerkraut producer (4) __ __ __ __ __ __ __ __ __ __ __
26. Causes ergot poisoning (3) __ __ __ __ __ __ __ __ __
27. Produces potent exotoxins (4) __ __ __ __ __ __ __ __ __ __ __
28. Green rot in foods (4) __ __ __ __ __ __ __ __ __ __ __
29. Used to cure cucumbers (5) __ __ __ __ __ __ __ __ __ __ __ __
30. Aflatoxins in grains (4) __ __ __ __ __ __ __ __ __ __ __

STEP B: CONCEPT REVIEW

31. Explain what is meant by **food spoilage** and identify the source for most food spoilage. (**Key Concept 1**)
32. Describe the conditions that promote food spoilage by microorganisms. (**Key Concept 1**)
33. Identify the microbial products responsible for food spoilage. (**Key Concept 1**)
34. Assess how meats and fish become contaminated. (**Key Concept 1**)
35. Describe the ways poultry and eggs become contaminated with microorganisms. (**Key Concept 1**)
36. Determine how breads and bakery products can support microbial growth. (**Key Concept 1**)
37. Identify the two types of fungi that can spoil grain by producing toxins. (**Key Concept 1**)
38. Discuss how milk and dairy products become soured or spoiled. (**Key Concept 1**)
39. Describe the methods used to preserve milk and other beverages. (**Key Concept 2**)
40. Assess the effectiveness of low temperatures for preserving foods and beverages. (**Key Concept 2**)
41. Summarize how conventional drying and **freeze-drying** preserve foods. (**Key Concept 2**)
42. Identify the chemical preservatives used in foods and beverages and describe how they contribute to preservation. (**Key Concept 2**)
43. Discuss how radiation can preserve foods and contribute to their shelf life. (**Key Concept 2**)
44. Contrast between a **foodborne infection** and a **foodborne intoxication**. (**Key Concept 2**)
45. Assess the importance of the **HACCP** system to food safety. (**Key Concept 2**)
46. Summarize how fermentation can act as a food preservation mechanism and name several common fermented foods. (**Key Concept 3**)
47. Discuss the role of microorganisms to the cheese manufacture process. (**Key Concept 3**)

STEP C: APPLICATIONS AND PROBLEMS

48. Having read this chapter, you decide to outfit your refrigerator with ultraviolet lights. What are you trying to accomplish?

49. Chicken and salad are two items on the dinner menu at home, and you are put in charge of preparing both. You have a cutting board and knife for slicing up the salad items and for cutting the chicken into pieces. Which task should you perform first? Why? What other precautions might you take to ensure that the meal is not remembered for the wrong reason?

50. One day you discovered an unopened container of sour cream that had been hidden in the back of the refrigerator for some 3 months. The sour cream appeared satisfactory, and there was no unusual smell. You proceeded to spoon it onto a baked potato and dig in. What factors might have contributed to the sour cream's preservation so long after the expiration date?

51. You and a friend go to an orchard to pick apples with the intention of pressing them in your new cider mill. However, you are aware of the recent outbreaks of infection with *E. coli* 0157:H7 that were traced to fresh apple cider. What precautions can you take to ensure a "healthy" experience?

STEP D: QUESTIONS FOR THOUGHT AND DISCUSSION

52. Suppose you had the choice of purchasing "yogurt made with pasteurized milk" or "pasteurized yogurt." Which would you choose? Why? What are the "active cultures" in a cup of yogurt?

53. Why might blue (Stilton) cheese pose a possible threat to someone who has an acute allergy to penicillin?

54. Which principles of preservation ensure that each of the following remains uncontaminated on the pantry shelf: vinegar, olive oil, brown sugar, tea bags, spaghetti, hot cocoa mix, pancake syrup, soy sauce, and rice?

55. On Saturday, a man buys a steak and a pound of liver and places them in the refrigerator. On Monday, he must decide which to cook for dinner. Microbiologically, which is the better choice? Why?

56. Foods from tropical nations such as Mexico tend to be very spicy, with lots of hot peppers, spices, garlic, and lemon juice. By contrast, foods from cooler countries such as Norway and Sweden tend to be much less spicy. Why do you think this pattern has evolved over the ages?

57. A standard set of recommendations and regulations exists for individuals who work in restaurants and cook food for customers (i.e., food handlers). However, very few regulations exist for individuals who pick fruits or vegetables in the fields. What regulations would you recommend for such workers?

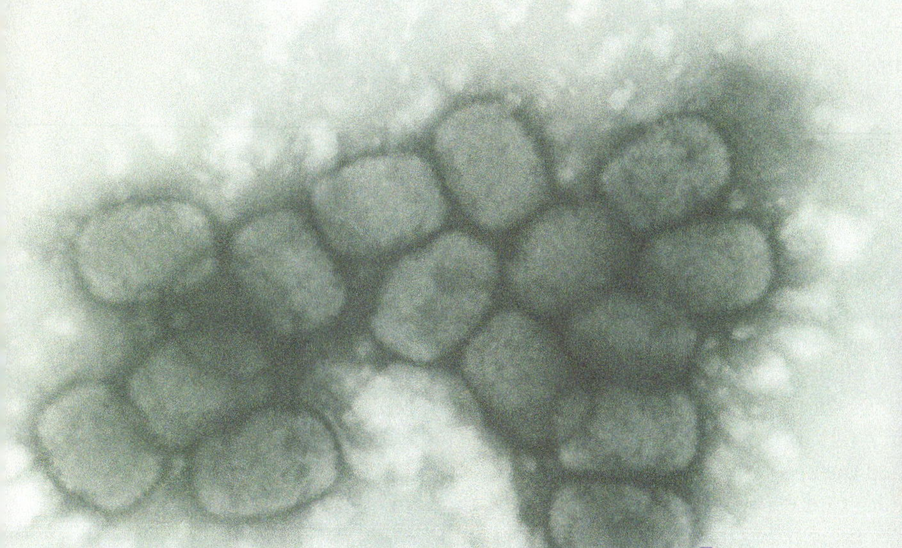

Environmental Microbiology

26

CHAPTER PREVIEW

Access to safe water is a fundamental human need and, therefore, a basic human right. Contaminated water jeopardizes both the physical and social health of all people. It is an affront to human dignity.
—Kofi Annan, former United Nations Secretary-General

In the early 1800s, the steam engine and its product, the Industrial Revolution, brought crowds of rural inhabitants to European cities. To accommodate the rising tide, row houses and apartment blocks were hastily erected, and owners of existing houses took in tenants. Not surprisingly, the bills of mortality from typhoid fever, cholera, tuberculosis, dysentery, and other diseases mounted in alarming proportions.

As the death rates rose, a few activists spoke up for reform. Among them was an English lawyer and journalist named Edwin Chadwick. Chadwick subscribed to the then-novel idea that humans could shape their environment and could eliminate diseases of filth by doing away with filth. In 1842, he published a landmark report suggesting that poverty-stricken laborers suffered a far higher incidence of disease than people from middle or upper classes. Chadwick attributed the difference to the abominable living conditions of workers, and he believed most of their diseases were preventable. His report established the basis for the Great Sanitary Movement, a wave of reform that began in Europe.

Chadwick was not a medical man, but his ideas captured the imagination of both scientists and social reformers. He proposed that sewers be constructed using smooth ceramic pipes and then enough water be flushed through the system to carry waste to some distant depository. To work efficiently, the system required the installation of new water and sewer pipes, the development of powerful pumps to bring water into homes, and the elimination of older sewage systems. The cost would be formidable.

Image courtesy of Dr. Fred Murphy/CDC.

Chadwick's vision eventually came to reality, but it might have taken decades longer without the intervention of cholera. In 1849, a cholera epidemic broke out in London and terrified so many people that public opinion began to form in favor of Chadwick's proposal. Another epidemic occurred in 1853, during which John Snow was able to trace the transmission of the disease back to a contaminating water pump. In both outbreaks, the disease reached the affluent as well as the poor, and the mortality rate exceeded 50%. Construction of the sewer system began shortly thereafter.

The proverbial "icing on the cake" came in 1892 when a devastating epidemic of cholera erupted in Hamburg, Germany. For the most part, Hamburg drew its water directly from the polluted Elbe River. Adjacent to Hamburg lay Altona, a city where the German government had previously installed a water filtration plant. Altona remained free of cholera. The contrast was sharpened further by a street that divided Hamburg and Altona. On the Hamburg side of the street, multiple cases of cholera broke out; across the street, none occurred. Chadwick and his fellow sanitarians could not have imagined a more clear-cut demonstration of the importance of water purification and sewage treatment.

Adequate safe water remains a serious problem in much of the world today (FIGURE 26.1). In fact, 1 billion people lack access to safe and sufficient drinking water; 4 billion people lack adequate sanitation; and 4 million deaths occur each year—primarily the poor and children—from insufficient and unsafe drinking water.

To address this and other health issues, in 2000 the leaders of 189 countries adopted the United Nations Millennium Declaration, which is a set of goals to solve many global environmental problems by 2015. One of these goals is to reduce by 50% the proportion of people without sustainable access to safe drinking water and sanitation.

Dealing with water pollution and treating sewage are but two of the myriad activities involving microbiology in public health. Other environmental issues to address involve the biogeochemical cycles, whose microorganisms are critical to the recycling of nitrogen, sulfur, and carbon. Thus, this chapter focuses on environmental microbiology and the study of microorganisms as they affect the natural environment.

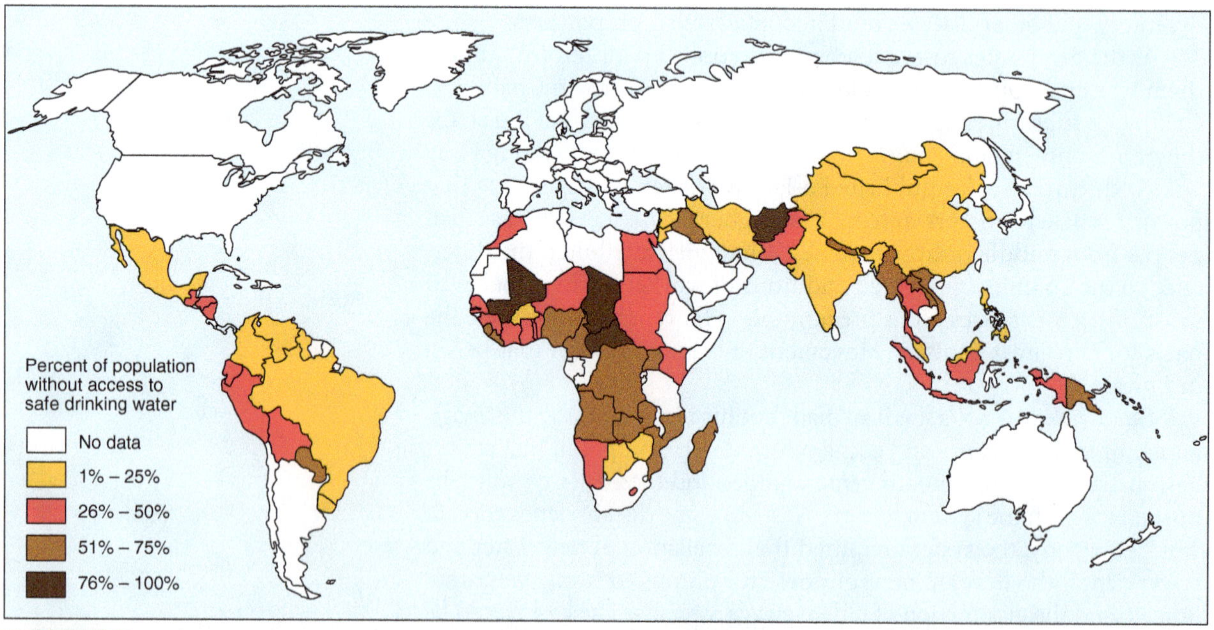

FIGURE 26.1 Populations Lacking Access to Safe Drinking Water. In 2012, more than 1 billion people lacked access to clean and sufficient drinking water. Modified from *The World's Water 2011–2012* by Peter H. Gleick. © 2011 Island Press. Reproduced by permission of Island Press, Washington, D.C. »» Provide several reasons why many developing nations lack clean water?

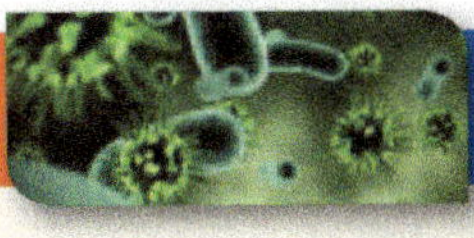

Chapter Challenge

Every year many lakes, reservoirs, and ponds undergo visible changes during the warm summer months. The usually clear, transparent water suddenly becomes soupy in appearance and may appear turquoise, greenish-yellow, or even red in color. One of the major organisms responsible for these changes is some members of the phylum Cyanobacteria. When these organisms, which occur naturally in surface waters, grow abundantly and in large masses called a bloom in a municipal water supply they may impart unpleasant tastes and odors to the water and in recreational waters can pose a serious health threat to other animals and humans as some common cyanobacteria produce potent enterotoxins and neurotoxins. On the other hand, cyanobacteria are key players in global carbon fixation and, as such, through photosynthesis are responsible for almost 50% of the atmospheric oxygen we and other animals breathe and all aerobic organisms use. Let's use the cyanobacteria as examples of how microbes affect the environment—the topic of this chapter.

KEY CONCEPT 26.1 Water Pollution Includes Biological Changes Harmful to Water Quality

For purposes of simplification, scientists classify water into two major types: groundwater and surface water. **Groundwater** originates from deep wells and subterranean springs, and, because of the filtering action of soil, deep sand, and rock, it is virtually free of microorganisms. As the water flows up along channels, physical, chemical, and biological contaminants may enter it and alter its quality. **Surface water** is found in lakes, streams, and shallow wells. Its microbial population may reflect the air through which rain has passed, agricultural or livestock operations, wildlife, or the sewage-treatment facility located along a riverbank.

Therefore, water is considered contaminated when it contains a chemical or biological poison, or an infectious agent. In polluted water, the same conditions apply, but the poison or agent is obvious, as it has an unpleasant taste, smell, or appearance. **Potability** (*pot* = "drink") by contrast, refers to water that is safe to drink.

Unpolluted and Polluted Water Contain Different Microbial Populations

A body of unpolluted water, such as a mountain lake or stream, is usually low in organic nutrients, and thus, only a limited number of bacterial species are present, perhaps a few thousand per milliliter. Most are soil organisms that have run off into the water during a rainfall. An example is the actinomycetes, a group of mold-like bacterial species responsible for the musty odor to soil. Other inhabitants include yeasts, along with bacterial and mold spores. Cellulose digesters such as members of the genus *Cellulomonas* are also found. These bacterial species digest cellulose in plant cell walls. Autotrophic bacterial cells are common, and free-living protozoa such as *Paramecium*, *Amoeba*, and *Euglena* abound.

A polluted body of water, such as a lake or river, presents a totally different picture (FIGURE 26.2A). The water contains large amounts of organic matter from sewage, feces, and industrial sources, and the population of microorganisms is usually heterotrophic. Such polluted water usually contains **coliform bacteria**, a group of gram-negative, non–spore-forming bacilli usually found in the human intestine (FIGURE 26.2B). Coliform bacteria, which ferment lactose to acid and gas, include *Escherichia coli* and species of *Enterobacter*. Noncoliform bacteria also common in polluted water include *Streptococcus*, *Proteus*, and *Pseudomonas* species.

In polluted water, microorganisms contribute to a chain of events that drastically alters the ecology of the environment (FIGURE 26.3). When phosphates accumulate in the water, algae reproduce rapidly. These **algal blooms** supply nutrients to other microorganisms, which multiply rapidly and use up the available oxygen. Soon, protozoa, small fish, crustaceans, and plants die and accumulate

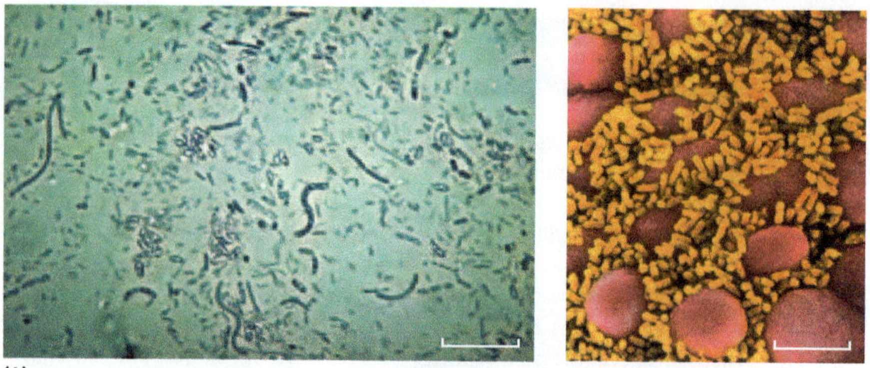

(A) (B)

FIGURE 26.2 **Microorganisms in Water Environments.** (**A**) A light micrograph of bacteria in polluted water. Many different bacterial shapes can be seen. (Bar = 10 μm.) © Eric Grave/Photo Researchers, Inc. (**B**) A false-color scanning electron micrograph of coliform bacteria. These *E. coli* bacteria on the intestinal lining can be defecated into the water. (Bar = 4 μm.) © EM Unit, VLA/ Photo Researchers, Inc. »» What bacterial shapes can you identify in the light micrograph?

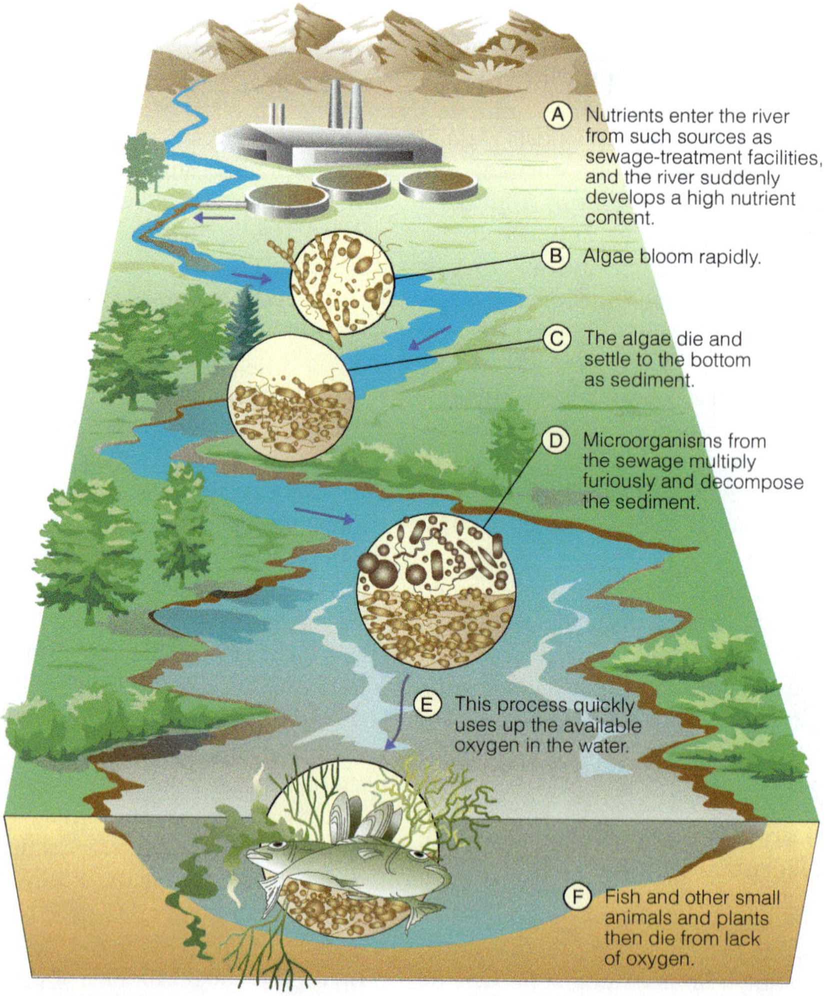

FIGURE 26.3 **The Death of a River.** The introduction of nutrients can lead to an algal bloom from which other microorganisms derive nutrients. This metabolism can quickly deplete the oxygen in the water, leading to the death of larger organisms, such as fish. »» Why would a lack of oxygen kill the fish?

on the bottom. Anaerobic bacterial species such as *Desulfovibrio* and *Clostridium* then thrive in the mud and produce gases, giving the water a stench reminiscent of rotten eggs.

The marine environment of the oceans illustrates another view of microbial populations in water. In the high salt concentration of ocean water, salt-tolerant organisms, called **halophiles**, survive. In addition, the organisms are **psychrophiles** because it is very cold below the surface. Those at the bottom must withstand great pressure and are therefore **barophiles**, or pressure tolerant. The organisms in these environments pose no threat to humans because they cannot grow in the body tissues.

Marine microorganisms are vital to ecological cycles because they form the foundations of many food chains. For example, marine algae such as **diatoms** (FIGURE 26.4A) and organisms such as the **dinoflagellates** capture the sun's energy and, using carbon dioxide, convert the energy to chemical energy in carbohydrates. The microorganisms then are consumed by other animals in the food chain. Dinoflagellates frequently make the news when certain species of *Gonyaulax* and *Gymnodinium* produce blooms called red tides (FIGURE 26.4B). The toxins produced by these organisms can kill many fish species.

Most marine microorganisms are found along the shoreline, or **littoral zone**, because this is where nutrients are plentiful. Certain unusual types of microorganisms have been found on the ocean floor in the **benthic zone** and even at the bottom of 10-kilometer-deep trenches, called the **abyssal zone**. Two types of protozoa, the foraminiferans and radiolarians, are of special interest to oil companies because these protozoa were dominant species during formation of the oil fields, and their fossils serve as markers for oil-bearing layers of rock.

There Are Three Types of Water Pollution

Water is vital to such industries as food processing, meat packing, and paper manufacturing. Water also is used extensively in pharmaceutical plants and mines, and as a coolant in power-generating units. It irrigates agricultural lands and provides the focus for many recreational facilities. Uses such as these, however, commonly add to the contamination and pollution of water.

Physical pollution of water occurs when particulate matter such as sand or soil makes the water cloudy, or when cyanobacteria bloom during midsummer and their remains give water

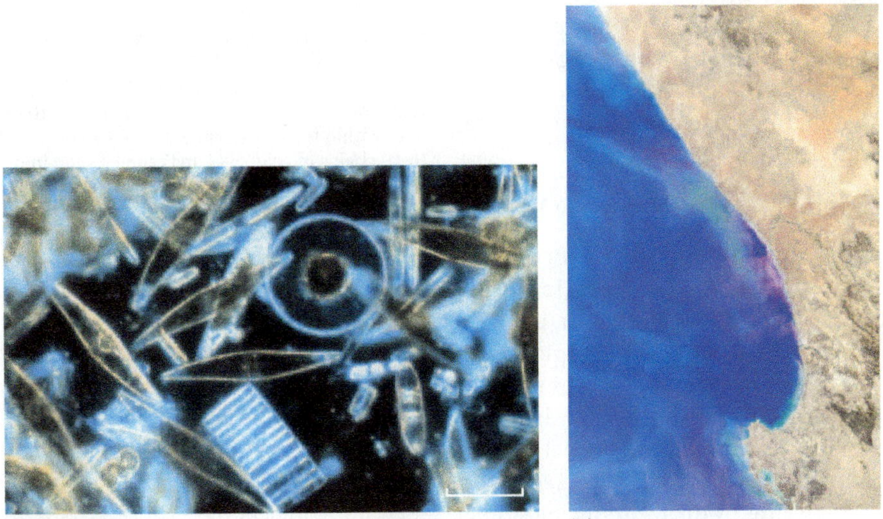

(A) (B)

FIGURE 26.4 **Marine Microorganisms.** (**A**) A dark field micrograph of a collection of diatoms. Note the broad variety of shapes and sizes in this group. Diatoms trap the sun's energy in photosynthesis and use it to form carbohydrates that are passed on to other marine organisms as food. (Bar = 25 μm.) Courtesy of Professor Gordon T. Taylor, Stony Brook University/NSF Polar Programs/NOAA. (**B**) Blooms of dinoflagellates, such as this one off the coast of Africa can have such high rates of metabolism that they release substantial amounts of toxins that poison fish. Courtesy of GSFC/LaRC/JPL, MISR Team/NASA. »» What causes an algal bloom such as in (B)?

the consistency of pea soup. **Chemical pollution** results from the introduction of inorganic and organic waste to the water. For example, water passing out of a mine contains large amounts of copper or iron. Other chemical pollutants in water include phosphates and nitrates from laundry detergents as well as acids like sulfuric acid.

The third type of pollution, **biological pollution**, is the main concern of our discussion. This type of pollution develops from microorganisms entering water from human waste, food-processing and meat-packing plants, medical facilities, and similar sources. Normally, water can handle biological material because heterotrophic microorganisms digest organic matter to carbon dioxide, water, and useful ions (phosphates, nitrates, and sulfates). With the rapid movement of the water, aeration is constant, and waste is soon diluted and eliminated. However, when water stagnates or is overloaded with waste, it becomes polluted.

A critical measurement in polluted water is the **biochemical oxygen demand** (**BOD**). This refers to the amount of oxygen microorganisms require to decompose the organic matter in water. As the number of microorganisms increases, the demand for oxygen increases proportionally. In the laboratory, the BOD is determined by measuring the dissolved oxygen content of water immediately after collection and then after incubation at 20°C for 5 days. The difference in oxygen content represents the amount used up by microorganisms in the water sample. Results are generally expressed as parts per million (ppm), with a BOD of several hundred ppm usually considered high.

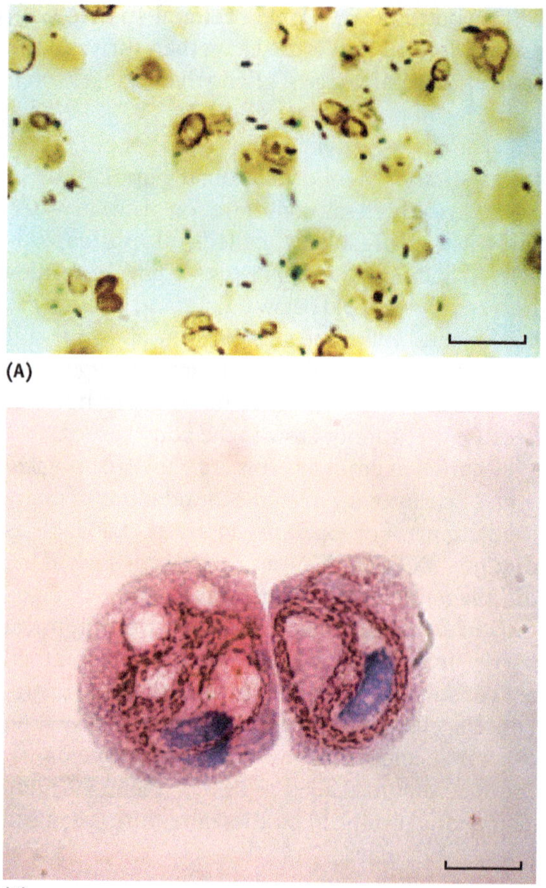

(A)

(B)

FIGURE 26.5 ***Legionella* in the Environment.** *Legionella pneumophila* is the gram-negative rod responsible for Legionnaires' disease. (**A**) It has infected the respiratory passages and cells in this biopsied lung tissue specimen (dark rods). Courtesy of CDC. (**B**) In nature, the pathogen lives within the cytoplasm of the waterborne protist. Here, chains of cells (dark red) can be seen in cells of *Tetrahymena*. (Bars = 10 µm.) Courtesy of Don Howard/CDC.
»» Because *Legionella* can be transmitted in airborne water droplets from a contaminated air conditioner, how would the bacterial cells survive while in the air conditioning system?

Diseases Can Be Transmitted by Water

Water consumption may be the vehicle for transfer of a broad variety of intestinal diseases, including bacterial diseases such as typhoid fever, cholera, shigellosis, and others. Waterborne epidemics, however, are rare because of continual surveillance. Many illnesses transmitted by drinking water are due to less familiar bacterial genera such as *Campylobacter*, *Yersinia*, and *Legionella* (FIGURE 26.5) as well as by toxin-producing strains of *E. coli*. Septicemia, necrotizing fasciitis, and gangrene may develop from an initial infection. An emerging pathogen associated with contaminated water is *Vibrio vulnificus*, a gram-negative bacterial species that can cause serious intestinal illness and septicemia in individuals with preexisting liver disease or compromised immune systems. Raw oyster consumption has been implicated in many deaths due to *V. vulnificus*.

Several bacterial species in marine waters also contribute to wound infections when contaminated seawater contacts the exposed tissue. Among the marine pathogens is *Erysipelothrix rhusiopathiae*, a gram-negative pleomorphic rod. Infections in humans, called **erysipeloid**, usually occur on the hands or feet and are accompanied by a bright red, well-demarcated lesion that burns or itches. *Mycobacterium marinum*, an acid-fast rod related to the tubercle bacillus, causes a small papular lesion or **granuloma**, at the wound site, particularly near the knuckles of the hand. The infection may spread via the lymphatic system. Fish handlers, seafood workers, and aquatic sports enthusiasts may suffer infection from these organisms, and antibiotics are used to limit the growth of both (MicroFocus 26.1). *V. vulnificus* can cause wound infections involving gangrene and necrotizing fasciitis.

Viral diseases transmitted by water include hepatitis A, rotavirus disease, gastroenteritis due to Coxsackie viruses or noroviruses, and in rare instances, the polio virus. These diseases are generally related to fecal contamination of water.

Water may be a vehicle for the transfer of *Entamoeba histolytica*, *Giardia lamblia*, and *Cryptosporidium*. The dinoflagellate *Gonyaulax catanella* produces a toxin causing muscular paralysis and death from asphyxiation. The toxin is ingested from shellfish that feed on the dinoflagellate. Another dinoflagellate, *Gambierdiscus toxicus* is consumed by small fish that concentrate the toxin and pass it to larger fish, such as sea bass and red snapper. Human consumption of the fish leads to neurological and muscular intoxication and a condition called **ciguatera** fish poisoning (from *cigua* for "poisonous snail," originally thought to be the cause).

CONCEPT AND REASONING CHECKS 1

a. Explain why pathogens tend to be found in polluted water.

b. Identify the three forms of water pollution and describe how they differ.

c. Summarize the types of infections caused by marine microorganisms.

MICROFOCUS 26.1: Environmental Microbiology

Fish Handler's Disease

It was the start of the rockfish (striped bass) season in 2004, and recreational fishermen were flocking to the Chesapeake Bay. The previous year brought in a record catch of the popular game fish, so everyone was looking forward to another great year. What wasn't publicized was the appearance of fish handler's disease on some fishermen.

Fish handler's disease usually develops on the skin of the hands or feet. Although there are several infections commonly referred to as fish handler's disease, abrasions on the skin most commonly become infected with *Mycobacterium marinum* present in polluted water or in infected fish. Human infections produce nodular lesions, which, although not life-threatening, can spread to the wrist and elbow, producing arthritis-like joint problems. The disease also occurs in aquarium shop personnel who clean infected aquaria and do not wear gloves (see figure).

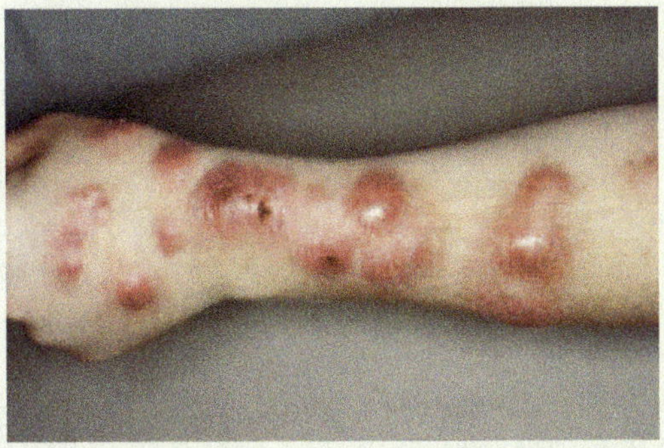

Nodules resulting from infection by *Mycobacterium marinum*. Courtesy of Chau Nguyen, M.D., Memorial University of Newfoundland.

In the 1980s, overfishing in the Chesapeake Bay had so decimated the rockfish number that a moratorium on fishing was imposed. Then by the mid-1990s, the population had recovered and the ban was lifted. However, the polluted nature of the Chesapeake Bay led to the identification of diseased rockfish. Now, it is estimated that 75% of the fish in the bay are diseased with *M. marinum*.

Fisheries biologists and researchers are not sure how the fish become infected. They know most infections occur in the Chesapeake Bay, which is where the fish breed. Pollution of the bay certainly plays a major role. Fish may look healthy, but *M. marinum* infects the spleen and can spread to other organs. Sores develop and the fish dies.

Because fish handler's disease is not a reportable disease, exactly how many humans become infected is unclear. In 2004, there were 46 reported cases. More cases appear in the summer and fall when food for rockfish becomes more scarce and heat degrades the bay's water quality. Such stresses produce more diseased fish.

Researchers have found 10 different strains of *M. marinum* in the rockfish. Two of these strains are newly identified, so their effects on humans remain to be determined.

Chapter Challenge **A**

As mentioned in the challenge opener, algal blooms composed of cyanobacterial species usually occur in the warm summer months.

QUESTION A:

Propose several reasons (causes) to explain the sudden explosion in numbers of cyanobacteria during a bloom. In a natural body of water, like a lake, what could be the outcomes of such a bloom?

Answers can be found on the Student Companion Website in **Appendix D.**

KEY CONCEPT 26.2 Proper Treatment of Water and Sewage Ensures Safe Drinking Water

Some years ago, healthcare workers in Africa asked villagers to name their single greatest need. The answer was almost unanimous: "Water." A startling survey by the World Health Organization (WHO) indicates that three of every four humans alive today do not have enough water to drink, or, if water is available, the supplies are contaminated (see chapter opener and quote).

Water is unfit to drink when it contains human sewage, animal waste, or disease-causing microorganisms. However, the situation can be reversed through the proper management of water resources. Water-purification procedures prevent pathogenic microorganisms from reaching the body, while sewage-treatment processes remove pathogens from body waste products. In this section, we will examine how these are accomplished.

Water Purification Is a Three-Step Process

Three basic steps are included in the preparation of water for drinking: sedimentation, filtration, and chlorination FIGURE 26.6. As we go through the steps, realize that the treatment process does not necessarily produce sterile water, but rather clean, potable water that is free of pathogens.

Sedimentation uses large reservoirs or settling tanks to remove leaves, particles of sand and gravel, and other materials from the soil. Chemicals such as aluminum sulfate (alum) are dropped as a powder onto water. The alum sticks together small particles in the water and forms jelly-like masses of coagulated material called **flocs**. In the process of **flocculation**, these flocs fall through the water and cling to organic particles and microorganisms, dragging a major portion to the bottom sediment. MicroFocus 26.2 describes a new, environmentally safe flocculant.

The **filtration** step is next. Although different types of filtering materials are available, most filters use a layer of sand and gravel to trap microorganisms. A slow sand filter, containing fine particles of sand a meter deep, is efficient for smaller-scale operations. Within the sand, a biofilm of microorganisms (bacterial, fungal, and protozoal cells) acts as a supplementary filter. This layer is called a **schmutzdecke** (German for "grime or filth cover"), or dirty layer. A slow sand filter may purify over 3 million gallons of water per acre per day. To clean the filter, the top layer is removed and replaced with fresh sand.

Some water treatment plants use membrane filter system, where the water flows through submerged hollow fibers with tiny pores. The water molecules can pass through the pores but the larger contaminants stay within the fibers and are filtered out. The filtered water is then pumped on to the next stage of treatment. Both forms of filtration eliminate more than 99% of the microorganisms and other very small particles from water. Often granular activated carbon, a black, sand-like material, is then used to remove natural organic matter. This step decreases taste and odors in the water.

The final step is **chlorination**. Water is disinfected with chlorine gas, which is an active oxidizing agent that reacts with any organic matter in water. It is important, therefore, to continue adding chlorine until a residue is present. A residue of 0.2 to 1.0 parts of chlorine per million (ppm) of water often is the standard

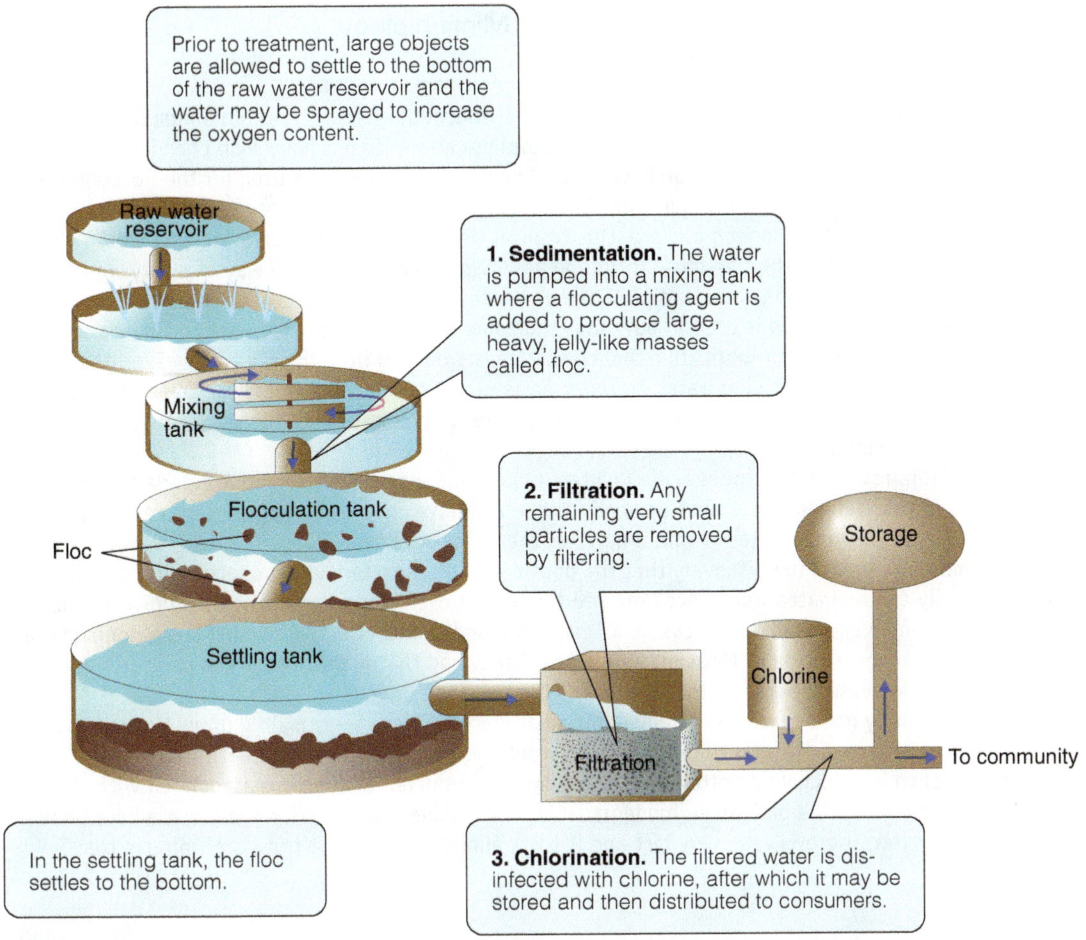

FIGURE 26.6 **Steps in the Purification of Municipal Water Supplies.** Water purification involves sedimentation, filtration, and chlorination. »» What would happen if the filtration step preceded the sedimentation step?

used in a water storage reservoir. Under these conditions, most remaining microorganisms die within 30 minutes.

Some communities also soften water by removing magnesium, calcium, and other salts. Softened water mixes more easily with soap, and soap curds do not form. Water also may be fluoridated to help prevent tooth decay. Scientists believe fluoride strengthens tooth enamel and makes the enamel more resistant to the acid produced by anaerobic bacterial species in the mouth.

Finally, in some areas of the United States, groundwater contains naturally or contaminating levels of arsenic. Removal requires pretreatment with ferric sulfate prior to the sedimentation and filtration steps.

Sewage Treatment Can Be a Multistep Process

Systems for the treatment of human waste range from the primitive outhouse, which is nothing more than a hole in the ground, to the sophisticated sewage-treatment facilities used by many large cities. All operate under the same basic principle: Water is separated from the waste, and the solid matter is broken down by microorganisms to simple compounds for return to the soil and water.

In many homes, human waste is emptied into underground **cesspools**, which are concrete cylindrical rings with pores in the wall. Water passes into the soil through these pores and pores on the bottom of the cesspool, where solid waste accumulates. Microorganisms, especially anaerobic bacterial

MICROFOCUS 26.2: Environmental Microbiology
Purifying Water with the "Miracle Tree"

In developed nations, water purification usually uses chemical powders such as aluminum sulfate or iron sulfate for the flocculation process. Many developing nations do not have such chemicals available, nor can they readily afford to purchase them. Could something else be used for the flocculation step in water purification? Yes! It's the "miracle tree."

Scientifically known as *Moringa oleifera*, this tropical tree was given the name "miracle tree" because of the many uses the tree has—including an environmentally friendly way to purify water.

M. oleifera survives in arid areas and produces elongated seedpods. The local people use the leaves and roots for food, the wood is used in building, and some parts of the tree are used in traditional medicines. High-grade oil for lubrication and cosmetics is produced from the seeds. However, the seeds also have another use—purifying water. In those parts of Africa and India where the trees grow, people grind up the seeds and add the powder to cloudy water to precipitate all the solid particles. Clean drinking water results.

Ian Marison, research director at École Polytechnique Fédérale de Lausanne in Switzerland, and his colleagues have examined the ground-up seed residue and have isolated a charged peptide. When they add this peptide to cloudy water, within 2 minutes the water goes from cloudy to clear.

Even more exciting is the discovery that the peptide also has bactericidal properties and can disinfect heavily contaminated water. According to Marison, it even works against some drug-resistant strains of *Staphylococcus, Streptococcus,* and *Legionella,* and is also effective against some nonbacterial waterborne pathogens. In fact, Marison's group has discovered that if they tweak the peptide's structure, they can increase the antimicrobial effect of the peptide.

This is certainly good news for the developing nations where the trees mainly grow. However, if commercially produced, the plant flocculant also would be useful to developed nations because the traditional chemical flocculants often have safety and environmental concerns associated with them.

As Marison says, "It's a biological, biodegradable, sustainable resource. *Moringa* grows where there is very little water; it grows very, very fast and it costs almost nothing." It truly is a "miracle tree."

species, digest the solid matter into soluble products that enter the soil and enrich it. Some hardware stores sell enzymes and dried *Bacillus subtilis* spores to accelerate digestion.

Some homes have a **septic tank**, an enclosed concrete box that collects waste from the house. Organic matter accumulates on the bottom of the tank, while water rises to the outlet pipe and flows to a distribution box. The water is then separated into pipes that empty into the surrounding soil. Because digested organic matter is not absorbed into the ground, the septic tank must be pumped out regularly.

Sewers are at least as old as the Cloaca Maxima ("Great Sewer") of Roman times. Until the mid-1800s, however, sewers were simply elongated cesspools with overflow pipes at one end. They collected filth and had to be pumped out regularly. Finally, in 1842, Edwin Chadwick's report raised the possibility that sewage spreads disease, and soon thereafter a movement (fueled by a cholera outbreak) sprang up to sanitize European cities.

Small towns collect sewage into large ponds called **oxidation lagoons**, where the sewage is left undisturbed for up to 3 months to allow natural digestion of the organic matter to occur. During that time, aerobic bacterial species digest organic matter in the water, while anaerobic organisms break down sedimented material. Under controlled conditions, the waste may be totally converted to simple salts such as carbonates, nitrates, phosphates, and sulfates. At the cycle's conclusion, the bacterial cells die naturally, the water clarifies, and the pond may be emptied into a nearby river or stream.

Large municipalities rely on a mechanized sewage-treatment facility to handle domestic wastewater, which contains massive amounts of waste and garbage (FIGURE 26.7). The process involves primary, secondary, and sometimes tertiary treatment.

Primary Treatment. In the first step, a screen is used to remove grit and large insoluble waste. Then, the raw sewage is piped into huge open

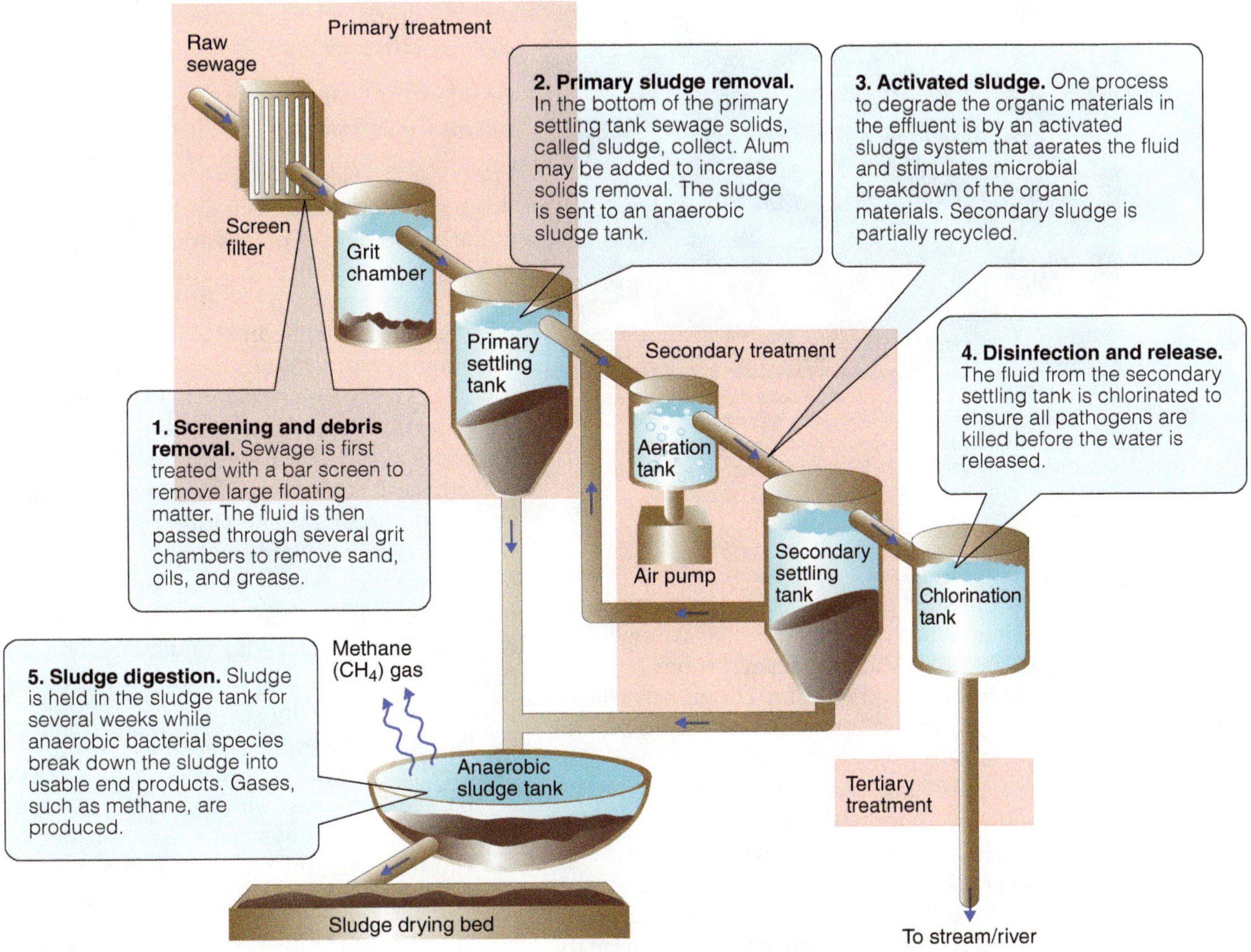

FIGURE 26.7 **A Sewage-Treatment Facility.** Sewage-treatment facilities use physical, chemical, and biological processes to treat and remove all water contaminants. »» Why must there be such an extensive sequence of steps to treat sewage?

sedimentation tanks for organic waste removal. This waste, called **sludge**, is passed into sludge tanks for further treatment. Flocculating materials, such as alum, may be added to the raw sewage to drag microorganisms and debris to the bottom.

Secondary Treatment. The purpose of secondary sewage treatment is to degrade the biological content of the fluid (effluent) coming from the primary treatment. This oxidation process is carried out by aerobic microorganisms as they metabolize the organic matter and produce inorganic end products (mainly CO_2 and H_2O). Two common secondary sewage treatment processes are the activated sludge system and trickling filters. They differ primarily in the way oxygen is supplied to the microorganisms and the rate at which organisms metabolize the organic matter.

In the activated sludge system, an aeration tank is supplied with oxygen so aerobic microorganisms can metabolize the organic matter in the primary effluent. One of the important bacterial species is *Zoogloea ramigera*, a gram-negative rod whose cells form a gelatinous matrix. Following 3 to 8 hours in the system, the active microorganisms (**activated sludge**) are separated from the liquid by sedimentation and the clarified liquid (secondary effluent) is further processed. A portion of the activated sludge is recycled back to the aeration tank to maintain a constant concentration of microbes. The remainder is removed and sent to the anaerobic sludge tank.

An alternative oxidation system is the "trickling filter" (also called a "biofilter"), which includes a basin or tower consisting of a bed of stones,

FIGURE 26.8 **A Trickling Filter.** In secondary sewage treatment, a trickling filter contains rotating pipes through which sewage waste is sprayed onto a rock or plastic bed of aerobic microbes. © Jonathan A. Meyers/Photo Researchers, Inc. »» Why is this system called a trickling filter?

gravel, molded plastic, or other porous substance that maintains a high surface-to-volume ratio (FIGURE 26.8). Microorganisms become attached to the bed and form a biofilm. The primary effluent containing the organic matter is sprayed over the bed. As the organic matter in the water percolates through the film, it is rapidly metabolized by the aerobic microbes. Oxygen is normally supplied to the film by the natural flow of air. Because the thickness of the biofilm increases as the microbes reproduce, fragments of the biofilm break off the bed and are combined with primary sludge in the anaerobic sludge tank.

The secondary effluent from secondary treatment processes may still contain a few pathogens. Therefore, the water is disinfected, usually by chlorination, before it is released into a stream, river, or ocean.

Finally, the primary and secondary sludge in the sludge tank is digested by anaerobic microbes; thus the sludge tank is very anaerobic. Methane-producing archaeal species predominate and give off methane and carbon dioxide gases. The methane gas may be used to provide the fuel to heat the sludge tank. The digested sludge is dried and used as a soil amendment or placed in landfills.

Tertiary Treatment. Sometimes further treatment of wastewater is needed as not all the water pollutants have been removed. Such tertiary sewage or advanced treatment systems are costly but may be essential for the removal of pesticides, fertilizers, and phosphate, which, as described earlier in this chapter, can pose problems if dumped into a lake or stream.

Biofilms Are Prevalent in the Environment

For decades, microbiologists have studied free-floating bacterial species growing in laboratory cultures as examples of how bacterial organisms live and behave in nature. In recent years, however, they have come to realize the importance of bacterial species living in biofilms. A **biofilm** is an immobilized population of bacterial species (or other microorganisms) living in a web of tangled polysaccharide fibers adhering to a surface. Examples of such environments are the surfaces of teeth, aquatic plants or animals, water pipes, and on stones of a trickling filter in secondary sewage treatment. Biofilms also are responsible for urinary tract infections and for contamination of medical devices, such as catheter tubes (FIGURE 26.9). Contact with a fluid environment ensures a plentiful supply of nutrients for the microbes of the biofilm.

Bacterial cells behave much differently in a biofilm than in a culture tube. Within a biofilm, aerobic species coexist with anaerobic species as they share passageways and interact metabolically. In this state, they form communities that store nutrients, benefit from each other's metabolic byproducts, and resist predators such as protozoa and bacteriophages. By contrast, **planktonic bacteria** live as individuals and do not enjoy these benefits. They tend to be of smaller size and have

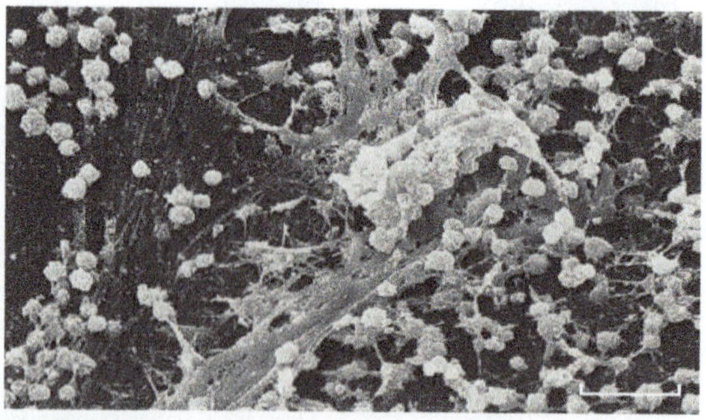

FIGURE 26.9 **Biofilm Contamination.** A scanning electron micrograph of a biofilm of *Staphylococcus aureus* on the inner surface of a needleless connector, which allows direct access to intravascular catheters. (Bar = 10 µm.) Courtesy of Dr. Rodney M. Donlan and Janice Carr/CDC. »» What do microorganisms gain by forming a biofilm?

a lower metabolic rate of respiration. Thus, the members of a biofilm are morphologically and physiologically distinct from their planktonic neighbors.

Microorganisms within a biofilm can chemically communicate with one another. Researchers have studied several bacterial species that form biofilms and have identified numerous extracellular signals. This so-called **quorum sensing** is discussed in another chapter.

Biofilms have both positive and negative effects. Among the benefits of biofilms are their uses in **bioremediation**, where biofilm populations are used to degrade toxic wastes and other synthetic products of industry. They also are employed in the oil industry to fill empty spaces after oil has been pumped out. In natural settings, they degrade organic compounds, thus retarding pollutant buildup.

On the negative side, concern continues to mount about the tendency of biofilms to form on contact lens surfaces, catheters, and medical implements (e.g., artificial hearts). Biofilms are important considerations in the development of dental caries and urinary tract infections. The concern is particularly acute because the bacterial species in biofilms resist phagocytosis by white blood cells, display enhanced resistance to antibiotics, and withstand the destructive action of disinfectants and antiseptics.

In the industrial setting, biofilms pose problems when their sulfate-reducing members anaerobically convert sulfur compounds to hydrogen sulfide. The latter corrodes water pipes, especially those that carry seawater (where sulfate abounds). When they contaminate computer chips, biofilms act as conductors and thereby interfere with electronic signals. Indeed, one researcher has called biofilms the "venereal disease of industry." Some scientists estimate that in nature, 99% of all microbial activities occur in biofilms.

The Bacteriological Analysis of Water Tests for Indicator Organisms

Many methods are available for detecting bacterial contamination of water, and various ones are selected according to the resources of the testing laboratory. Because it is impossible to test for all pathogenic microorganisms, water-quality bacteriologists have adopted the practice of testing for certain **indicator organisms**. If these bacterial species, normally found in the human intestinal tract, are present in a water sample, fecal contamination has probably taken place in the original body of water.

Among the most frequently used indicator organisms are the coliform bacteria. **Coliform bacteria** normally are found in the intestinal tracts (colon) of humans and many warm-blooded animals. They can survive for extensive periods of time in the environment, and they are relatively easy to cultivate in the laboratory. *E. coli* is the most important indicator organism within the group.

The **membrane filter** technique is a popular laboratory test in water microbiology because it is straightforward and can be used in the field. A technician places a specially designed collecting bottle in the water and takes a 100-mL sample (FIGURE 26.10). The water then is passed through a membrane filter, and the filter pad is transferred to a plate of bacteriological medium. Bacterial cells trapped in the filter will form colonies, and by counting the colonies, the technician can determine the original number of bacterial cells in the sample.

Another method for testing water is the **standard plate count** (SPC) technique. Samples of

FIGURE 26.10 **Collection of Water for Analysis.** A water sample is collected and potential indicator organisms are identified by the membrane filtration technique. Courtesy of Scott Bauer/USDA. »» What types of indicator organisms would indicate this pond is contaminated?

water are diluted in sterile buffer solution, and carefully measured amounts are transferred to culture dishes. Agar medium is added, and the plates are set aside at incubation temperatures. A count of the colonies multiplied by the reciprocal of the dilution (the dilution factor) yields the total number of bacterial cells per mL of the original sample.

A third test is a statistical evaluation called the **most probable number (MPN)** test. In this procedure, a technician inoculates water in 10-mL, 1-mL, and 0.1-mL amounts into lactose broth tubes. The tubes are incubated and coliform organisms are identified by their production of carbon dioxide gas. Referring to an MPN table, a statistical range of the number of coliform bacteria is determined by observing how many broth tubes showed gas. MicroInquiry 26 gives you a chance to carry out a standard qualitative water analysis.

CONCEPT AND REASONING CHECKS 2

a. Assess the need for three steps in water purification.
b. Summarize the steps in the treatment of sewage.
c. Assess the positive and negative effects of biofilms.
d. Describe the three lab tests for water quality.

Chapter Challenge B

Some cyanobacterial species potentially can cause problems for maintaining water quality, including taste and odor producing compounds and toxins that may be dangerous enough to poison animals and humans. Therefore, in places where community lakes or reservoirs are not only important recreational resources but also provide some percentage of a city's drinking water, cyanobacterial blooms and their products would pose a risk to a safe drinking water supply.

QUESTION B:
As a water utility manager, how would you ensure that the city's water purification process provides the citizens with safe drinking water?

Answers can be found on the Student Companion Website in **Appendix D.**

MICROINQUIRY 26
Doing a Standard Qualitative Water Analysis

One of the traditional methods of analyzing bacteriological water quality is through the **most probable number (MPN)** test. In this analysis, the detection of coliform bacteria (gram-negative enteric bacteria) involves up to three sets of tests, called the presumptive, confirmed, and completed tests.

The presumptive test is designed to detect coliform bacteria in a water sample. Of the enteric bacteria, only the coliform bacteria use lactose as a carbon and energy source. Therefore, a lactose broth medium represents a selective medium for coliforms. If coliforms are present, they ferment the lactose and produce acid and carbon dioxide gas. The presence of acid is seen by a change in the broth color (red to yellow) due to the presence of a pH indicator and gas production will be trapped in a small, inverted tube within the lactose broth tube (see figure).

Three sets of five lactose broth tubes each are used. Set 1 tubes contain 10 mL of double-strength lactose broth, while set 2 tubes contain 1 mL and set 3 tubes contain 0.1 mL of single-strength lactose broth. Each tube then is inoculated with a known volume of the water sample. If any of the tubes produce gas, you can presume that coliform bacteria are present in the original water sample.

If a presumptive test indicates the presence of coliforms, you can determine the MPN by noting how many of the tubes in each group have produced acid and gas. Then, using a standard MPN table, you can obtain a statistical estimate for the number of coliforms present (see table). For example, if three of the 10 mL tubes, two of the 1 mL tubes, and zero of the 0.1 mL tubes had positive results, the presumptive test would be read as 3-2-0, indicating approximately 14 coliforms per 100 mL of water (95% confidence that there were actually between 5.7 and 36 bacteria present).

Scenario
Working for the state health department, your job is to regularly analyze lake water from a public recreation area for coliform bacteria that would indicate fecal pollution. The health department needs to ensure that the water contains no appre-

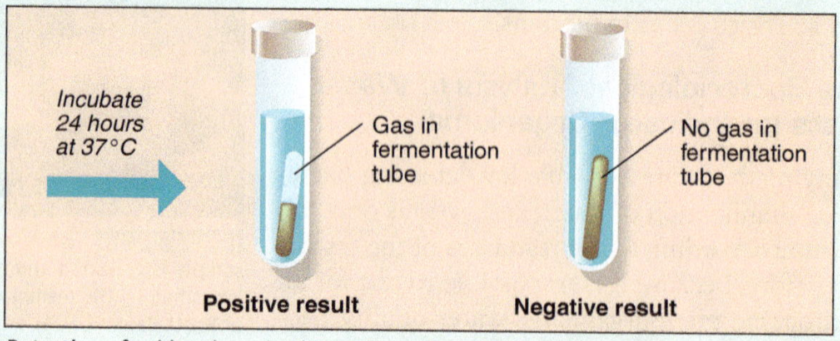

Detection of acid and gas in the presumptive test for water quality.

ciable number of coliform bacteria that could otherwise make people sick after swimming or playing in the water. Let's explore the presumptive test and then mention a few details about the confirmed and completed tests. Answers can be found on the Student Companion Website in **Appendix D.**

You have obtained six lake water samples from the public recreation area. You set up a presumptive test using 15 lactose broth tubes (five each with 10 mL, 1 mL, and 0.1 mL of broth) for each of the six water samples. Using sterile pipettes, you transfer 10 mL aliquots into each of the 10 mL tubes, 1 mL aliquots into the 1 mL tubes, and 0.1 mL aliquots into the 0.1 mL tubes. All of the tubes are incubated at 37°C for 24 hours.

26.1. Why do the 10 mL tubes contain double-strength lactose broth?

26.2. Why were sterile pipettes used? Is this absolutely necessary for this test?

Here are the results for the six samples:

Sample Number	Positive Tubes 10 mL	1 mL	0.1 mL
1	0	0	1
2	0	0	0
3	0	1	0
4	5	4	5
5	0	2	0
6	2	0	1

26.3. What would you conclude from these presumptive test results?

The confirmed test is designed to confirm the presence of coliform bacteria in positive or uncertain presumptive test results.

26.4. Why is it necessary to confirm a positive presumptive test?

In a confirmed test, a sample from each identified lactose broth tube is streaked onto an eosin-methylene blue (EMB) agar plate. This is a selective and differential growth medium. It is selective because it inhibits the growth of gram-positive (nonenteric) bacteria, and it is differential because of any enteric bacteria present, only colonies of *Escherichia coli*—the prime indicator of fecal pollution—produce dark centers with a green metallic sheen.

26.5. Which lake water samples need to be confirmed? What was your reasoning for selecting these samples?

If needed, a completed test can be run for any samples that gave a doubtful result in the confirmed test.

26.6. Suppose that only sample 4 gives a positive confirmed test; as the water analysis expert, what would be your recommendation to the state department of health regarding this recreational lake?

TABLE

The MPN Index for Combinations of Positive and Negative Presumptive Test Results*

Positive Tubes 10 mL	1 mL	0.1 mL	MPN Index	Confidence Limit Low	High
0	0	0	<1.8	—	6.8
0	0	1	1.8	0.09	6.8
0	1	0	1.8	0.09	6.9
0	2	0	3.7	0.7	10
2	0	0	4.5	0.79	15
2	0	1	6.8	1.8	15
2	1	0	6.8	1.8	17
2	1	1	9.2	3.4	22
2	2	0	9.3	3.4	22
2	3	0	12	4.1	26
5	4	0	130	36	400
5	4	1	170	58	400
5	4	2	220	70	440
5	4	3	280	100	710
5	4	5	430	150	1100

*Per 100 mL when five each of 10-mL, 1-mL, and 0.1-mL sample portions are used.

KEY CONCEPT 26.3 Microbes Are Indispensable for Recycling Major Chemical Elements

The thought of microorganisms often conjures a negative reaction because of their disease implication, and the contents of this chapter have undoubtedly supported that notion. It would be unwise, however, to neglect the positive role of microorganisms in the environment, because it is a substantial one. In fact, if all microbes suddenly left Earth, all life would eventually cease. Their role can be appreciated by briefly examining the place of microorganisms in three vital cycles of nature: the carbon, sulfur, and nitrogen cycles.

The Carbon Cycle Is Influenced by Microorganisms

Planet Earth is composed of numerous chemical elements, among which is a defined amount of carbon that is constantly recycled via the **carbon cycle** to allow the formation of the organic compounds that comprise all living organisms. **Carbon fixation** by photosynthetic organisms takes carbon in the form of carbon dioxide and converts it into carbohydrates using the sun's energy and chlorophyll pigments. The vast jungles of the world, the grassy plains of the temperate zones, and the cyanobacteria of the oceans show the results of this process. Investigating the Microbial World 26 looks at how cyanobacteria maximize the carbon fixation process.

Investigating the Microbial World 26

Fixated on Carbon

Among the most abundant organisms on land, sea, and lakes are the heterogeneous, photosynthetic members of the phylum Cyanobacteria. The marine members account for 25% of Earth's photosynthetic productivity and through the carbon cycle and photosynthesis, the cyanobacterial cells sequester some 25 gigatons of carbon annually, which is almost three times the carbon released worldwide each year from burning fossil fuels. To be so efficient, one would assume the cyanobacterial cells must be highly organized. But traditionally the prokaryotes have not been considered as highly structured organisms—that was the province of the eukaryotes.

- **OBSERVATIONS:** *Synechococcus elongatus* is one of the most common cyanobacteria in the oceans. The unicellular, rod-shaped cells are about 1.0 μm to 1.5 μm in length and are found in the brightly lit surface waters at concentrations of 1,000 to 200,000 cells per milliliter. Electron microscopy has identified several internal subcompartments, including highly structured carboxysomes that, by having a protein shell, isolate the first reactions of carbon fixation from the cytoplasm.

- **QUESTION:** ***How are the carboxysomes arranged in cells of*** **S. elongatus*?***

- **HYPOTHESIS:** The carboxysomes have a specific arrangement controlled by the cell's cytoskeleton. If so, then microscopy should detect a loss of arrangement of carboxysomes in mutant cells with a cytoskeleton defect.

- **EXPERIMENT:** A yellow fluorescent protein (fluoresces green) was used to tag the protein shell that builds carboxysomes in wild-type *S. elongatus* cells. A similar tag was used to identify carboxysomes in a mutant strain that has a partially disabled cell cytoskeleton.

- **RESULTS AND CONCLUSIONS:** (see figure)

  QUESTION 1: ***What do the results show?***

  QUESTION 2: ***Was the hypothesis supported? Explain.***

  QUESTION 3: ***Suggest a mechanism to organize the carboxysomes in the cell cytoplasm.***

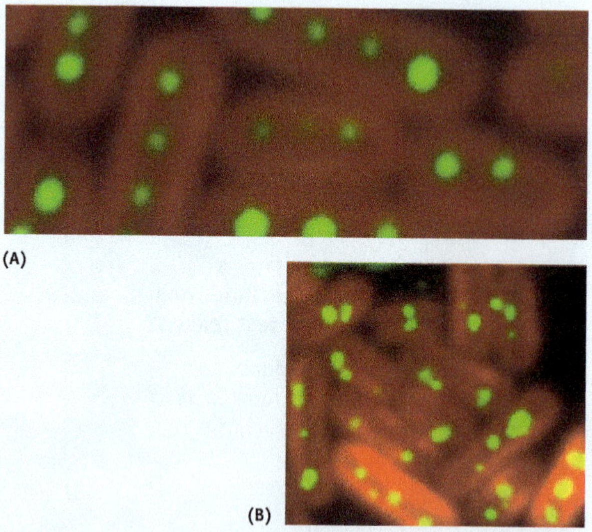

(A)

(B)

(**A**) Yellow fluorescent protein (green) was used to tag carboxysomes in wild-type cells. (**B**) A similar tag labeled carboxysomes in mutant cells. (Bars = 2 μm.) Courtesy of Pam Silver, Harvard Medical School.

Answers can be found on the Student Companion Website in **Appendix D.**

Adapted from: Savage, D. F., *et al.* (2010). *Science* **327**(5970): 1258–1261. © Tischenko Irina/ShutterStock, Inc.

Photosynthetic organisms, in turn, are consumed by grazing animals, fish, and humans, who use some of the carbohydrates for energy and growth, and convert much of the remainder into cellular materials. To be sure, some carbon dioxide is released back to the atmosphere in cellular respiration, but a major portion of the carbon is returned to the ground when the animal or plant dies (FIGURE 26.11). It is here that the microorganisms exert their influence, for they are the primary **decomposers** of dead organic matter. Working in their countless billions in the water and soil, many bacterial cells, fungi, and other microorganisms consume the organic substances and release atmospheric carbon dioxide for reuse by the plants. This activity results from the concerted action of a huge variety of microorganisms, each with its own nutritional pattern of protein, carbohydrate, or lipid digestion. Without microorganisms, the Earth would be a veritable garbage dump of animal waste, dead plants, and organic debris accumulating in implausible amounts.

Microorganisms accomplish a similar goal when they decompose manure and other natural materials and convert them to compost for crop fertilization. Although both conventional and organic agriculture use manure as part of regular farm soil fertilization programs, only certified organic farmers are required to have a plan detailing the methods for building soil fertility using raw or aged manure. Furthermore, certified organic farmers are prohibited from using raw manure for at least 60 days prior to harvesting crops for human consumption.

Microorganisms also break down the carbon-based chemicals produced by industrial processes, including herbicides, pesticides, and plastics. In addition, they produce methane, or natural gas, from organic matter and are probably responsible for the conversion of plants to petroleum and coal deep within the recesses of the Earth. Moreover, many microorganisms trap CO_2 from the atmosphere and form carbohydrates to supplement the results of photosynthesis. In these activities, the

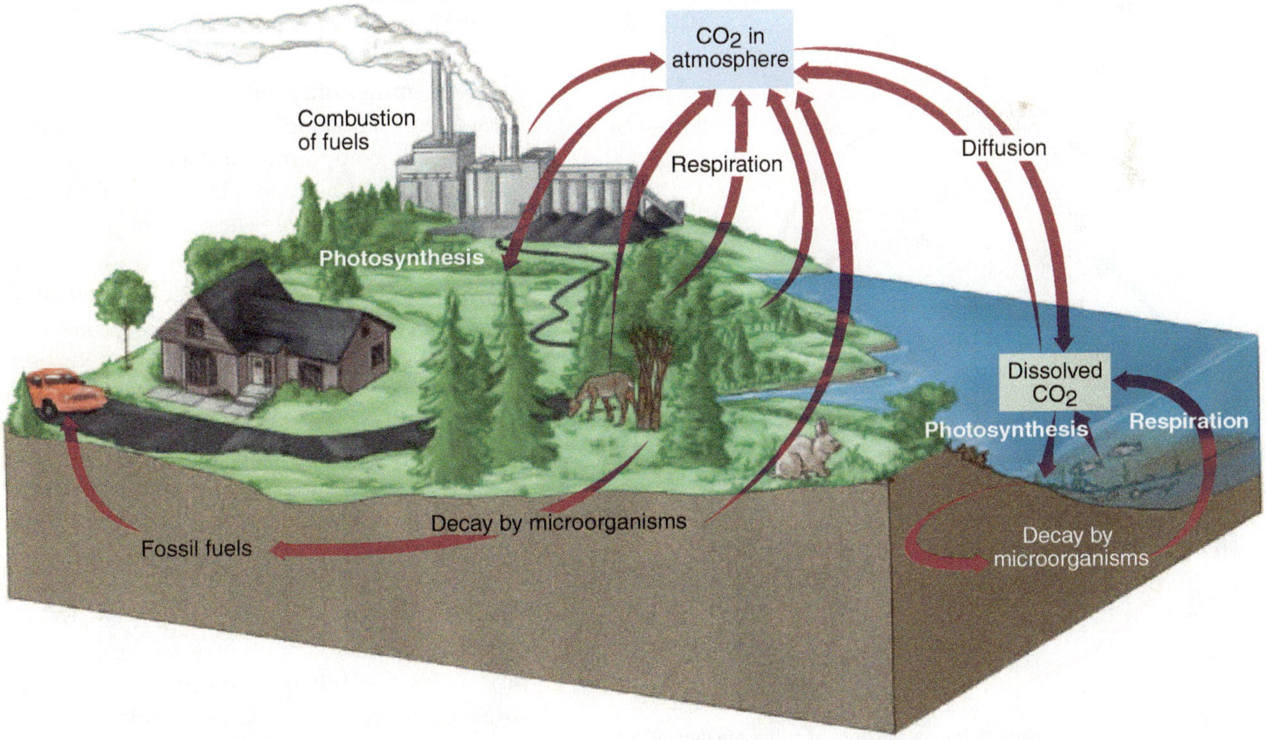

FIGURE 26.11 **A Simplified Carbon Cycle.** Photosynthesis represents the major method for incorporating carbon dioxide into organic matter, and cellular respiration accounts for its return to the atmosphere. »» Identify the roles played by microorganisms to decay in soil and ocean sediments.

microorganisms represent a fundamental underpinning of organic life.

The Sulfur Cycle Recycles Sulfate Molecules

The **sulfur cycle** may be defined in more specific terms than the carbon cycle. Sulfur is a key constituent of such amino acids as cysteine and methionine, which are important components of proteins. Proteins are deposited in water and soil as living organisms die, and bacterial and fungal species decompose the proteins and break down the sulfur-containing amino acids to yield various compounds, including hydrogen sulfide (FIGURE 26.12). Sulfur also may be released in the form of sulfate molecules commonly found in organic matter. Anaerobic genera, such as *Desulfovibrio,* subsequently convert sulfate to hydrogen sulfide through anaerobic respiration.

Other anaerobic conversions involve several bacterial genera, including members of the genera *Thiobacillus, Beggiatoa,* and *Thiothrix.* These bacterial genera release sulfur from hydrogen sulfide during their metabolism and convert it into sulfate. The sulfate now is available to plants and microbes, where it is assimilated into the sulfur-containing amino acids. Further assimilation by animals and humans completes the cycle.

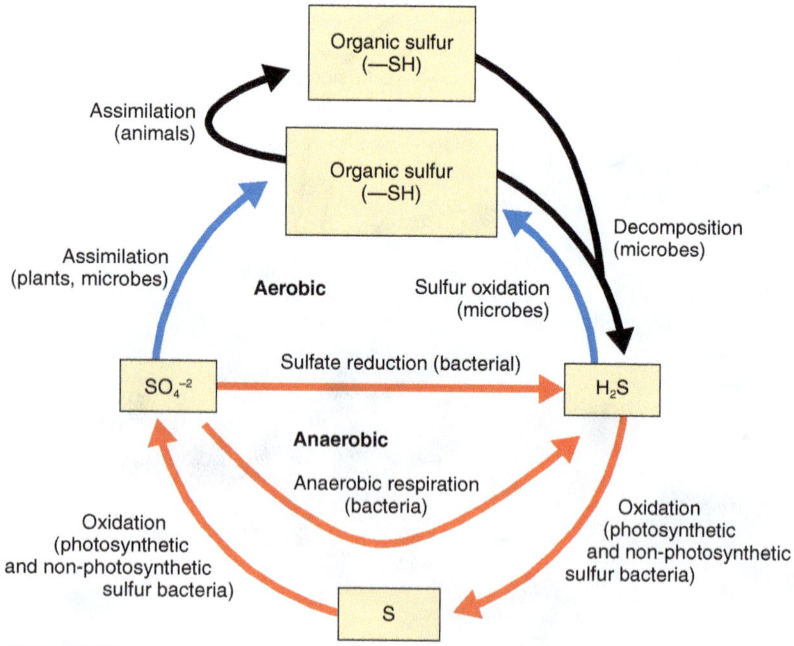

FIGURE 26.12 **The Sulfur Cycle.** Most of the conversions of sulfur are dependent on microbes in aerobic and anaerobic environments. The red lines indicate anaerobic conversions. »» What would happen if all microbes involved in the sulfur cycle disappeared?

The Nitrogen Cycle Is Dependent on Microorganisms

The cyclic transformation of nitrogen is of paramount importance to life on Earth. Nitrogen is an essential element in nucleic acids and amino acids. Although it is the most common gas in the atmosphere (about 80% of air), animals cannot use nitrogen in its gaseous form, nor can any but a few species of plants. The animals and plants thus require the assistance of microorganisms to trap the nitrogen. Among the first to recognize this relationship was Martinus Beijerinck, who did his work at the end of the 1800s (MicroFocus 26.3).

The **nitrogen cycle** begins with the deposit of dead plants and animals in the soil. In addition, nitrogen reaches the soil in urea contained in urine. A process of digestion and putrefaction by soil bacteria and other microorganisms follows, thus yielding a mixture of amino acids (FIGURE 26.13). Amino acids are broken down further by microbial metabolism, and the ammonia that accumulates may be used directly by plants.

Next, **mineralization** takes place. In this process, complex organic compounds are converted to inorganic compounds and additional ammonia. Much of the ammonia is converted to nitrite ions by *Nitrosomonas* species, a group of nitrifying, aerobic gram-negative rods. In the process, the bacterial cells obtain energy for their metabolic needs. The nitrite ions then are converted to nitrate ions by *Nitrobacter,* another group of aerobic gram-negative rods. *Nitrobacter* obtains energy from the process. Nitrate is a crossroads compound: It can be used by plants for their nutritional needs, or it can be liberated as atmospheric nitrogen by certain microorganisms, including the denitrifying bacteria.

For the nitrogen released to the atmosphere, a reverse trip back to living things is an absolute necessity for life to continue as we know it. The process is called **nitrogen fixation**. Once again, some bacterial and archaeal species in water and soil play a key role because they possess the enzyme systems to trap atmospheric nitrogen and convert it to compounds useful to plants. In

MICROFOCUS 26.3: History

Of Lumps and Bumps

The man from the Delft laboratory had an audacious proposal to the assembled farmers: "Don't plant your crops in the same field as last year," he said. "Leave the field alone for the next 2 years; let it lie fallow." The year was 1887. The man was Martinus Willem Beijerinck. The country was the Netherlands. And, because agricultural land was at a premium, the proposal was revolutionary.

Beijerinck was a local bacteriologist. While his medical colleagues were investigating the germ theory of disease and its implications, Beijerinck was out in the fields. He was observing that fields are very productive when the land had just been cleared and freshly planted. In fact, these fields yielded bountiful crops when the farmer is away for a couple of years. Now he thought he had the answer: great populations of bacterial organisms.

Beijerinck was an expert on plants, but he also had something most other botanists lacked—a solid background in chemistry. He believed nitrogen was essential for plant growth, but he had no idea how nitrogen bridges the gap between atmosphere and plant. Then it dawned on him that bacterial organisms might be the bridge. And the little lumps and bumps on tree roots were the key. Time and again he observed great hordes of bacterial cells in the little lumps and bumps ("nodules") on plant roots (see Figure 26.14A). He didn't see the nodules as often on tended crops, but they always seemed to be on wild plants growing in untended fields.

Beijerinck performed the laboratory experiments to strengthen his views. He took bacterial cells from the nodules and inoculated them into seedlings of various plants. In many cases, the plants developed nodules, and when he planted the seedlings, the nitrogen content of the soil rose dramatically. So, his advice in 1887 was straightforward: Leave the field alone for a spell; plant elsewhere; let the wild plants thrive; and when the field is finally planted, the crop yield will be worth the wait.

Indeed, he was right. Modern farmers know it is important to let a field lie fallow and "refresh" itself. To the macrobiologist, the lumps and bumps on the roots of wild plants are the important parts. To the microbiologist, it's what is inside that counts. Either way, we all benefit.

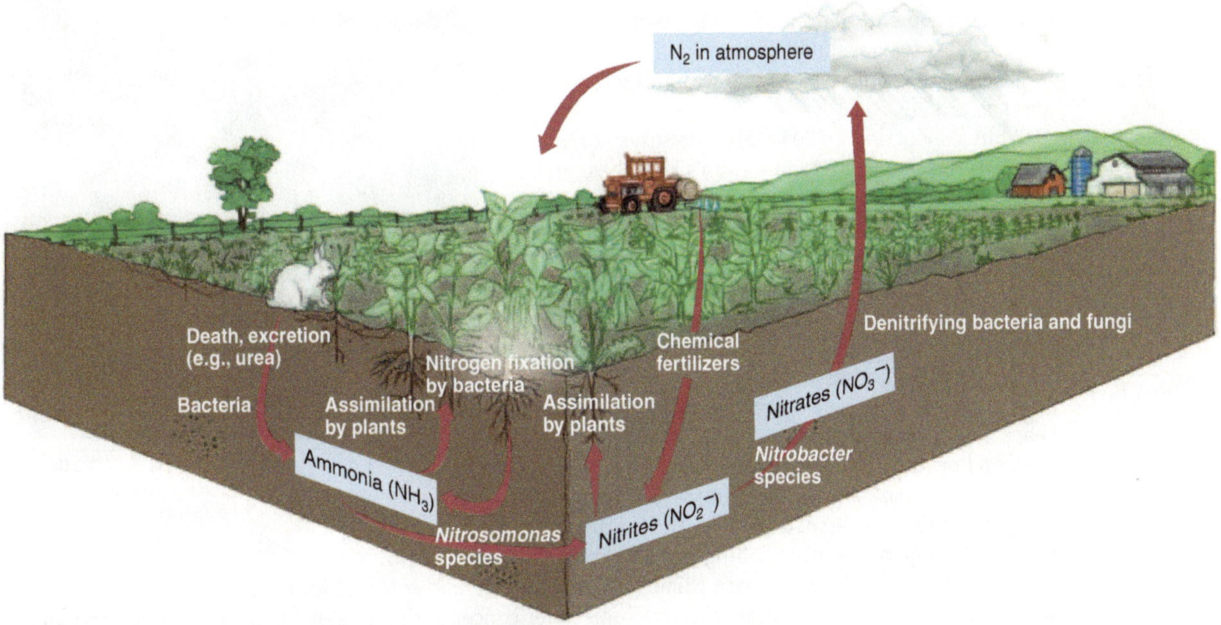

FIGURE 26.13 **A Simplified Nitrogen Cycle.** Plant and animal protein and metabolic wastes are decomposed by bacteria into ammonia. The ammonia may be used by plants, or it may be converted by *Nitrosomonas* and *Nitrobacter* species to nitrate, which also is used by plants. Some nitrate is broken down to atmospheric nitrogen. This nitrogen is returned to the leguminous plants by nitrogen-fixing microorganisms as nitrate, which is converted to ammonia. Animals consume the plants to nitrogen from the proteins plants contain.
»» If plants could directly carry out nitrogen fixation, would it make microorganisms unnecessary for the nitrogen cycle? Explain.

■ Mutualistic:
Refers to a symbiotic relationship between two organisms of different species that benefits both.

■ Symbiotic:
Refers to an association between two species that often is beneficial to both.

nitrogen fixation, gaseous nitrogen is converted into ammonia, which fertilizes plants.

Two general types of microorganisms are involved in nitrogen fixation: free-living species and symbiotic species. Free-living species include bacterial genera such as *Bacillus*, *Clostridium*, *Pseudomonas*, *Spirillum*, and *Azotobacter* as well as types of cyanobacteria and certain yeasts. Generally, the free-living species fix nitrogen during their growth cycles. The nitrogen-fixing ability of these species cannot be overemphasized.

Symbiotic species of nitrogen-fixing microorganisms live in association with plants that bear their seeds in pods. These plants, known as legumes, include peas, beans, soybeans, alfalfa, peanuts, and clover. Species of gram-negative rods known as *Rhizobium* infect the roots of the plants and live within swellings, or nodules, in the roots. Although complex factors are involved, the central theme of the **mutualistic** relationship is that *Rhizobium* fixes nitrogen and makes nitrogen compounds available to the plant while taking energy-rich carbon compounds in return. The bulk of the nitrogen compounds accumulates when *Rhizobium* cells die. Legumes then use the compounds to construct amino acids and, ultimately, protein. Animals consume the soybeans, alfalfa, and other legumes and convert plant protein to animal protein, thereby completing the cycle. In some cases, however, such *Rhizobium* nodules may not exist in the soil (MicroFocus 26.4).

Humans have long recognized that soil fertility can be maintained by both rotating crops and

MICROFOCUS 26.4: Environmental Microbiology

The Lows and Highs of Nitrogen Fixation

The textbook examples of nitrogen fixation have always emphasized the roles microbes play in the soils and in association with plant roots. Now it appears that at least in some tropical trees, the process can occur high up in the tree canopies.

The koa tree (see figure) is quite special to the Hawaiian people. It is from this tree that islanders made canoes for traveling between islands. As such, the tree became quite sacred and was protected by the god Kupulupulu.

A koa tree in Hawaii. Photo by Forest & Kim Starr.

While conducting a 2003 statewide survey of nitrogen fixation bacteria in Hawaii's koa trees, graduate student James Leary of the University of Hawaii, Manoa, came across something startling. He discovered that in a few koa trees, some nitrogen-fixing bacterial species reside not only in the soil in association with the tree's shallow roots, but also in the tree branches. Species of the bacterium *Rhizobium* were found quite high up in the branches.

When these trees were examined more closely, Leary found rhizobia nodules in clumps of decomposing wood or soil-like debris found both where the tree branches fork and on tree limbs. In these clumps, Leary found rhizobia associated with an amazing community of other microbes, including molds, other fungi, and lichens.

Further analysis indicated these clumps of detritus were better places for rhizobia to exist because there was a higher concentration of carbon and lower concentration of aluminum, which can interfere with nitrogen fixation. In fact, most soil-dwelling microbes were parasitic, while the canopy-dwelling ones were specialized for a mutualistic form of symbiosis.

(A) (B)

FIGURE 26.14 **Nitrogen Fixation.** (A) Nodules on the roots of a cowpea, a legume plant. Species of *Rhizobium* live within the nodules and fix nitrogen to nitrogen-containing compounds. When the bacterial cells die, the compounds are used by the legume to synthesize amino acids. © DS70/iStockphoto. (B) A false-color scanning electron micrograph of a *Rhizobium* species comprising a root nodule. (Bar = 10 µm.) © Medical-on-Line/Alamy Images. »» Compared to a leguminous plant with root nodules, what might a leguminous plant look like if it lacked nodules containing *Rhizobium*?

including a legume. The explanation lies in the ability of rhizobia to fix nitrogen within the nodules of legumes (FIGURE 26.14). So much nitrogen is captured, in fact, that the net amount of nitrogen in the soil actually increases after a crop of legumes has been grown. When cultivating legumes, there is no need to add nitrogen fertilizer to the soil. In addition, when crops such as clover or alfalfa are plowed under, they markedly enrich the soil's nitrogen content. Thus, humans are indebted to microorganisms for such edible plants as peas and beans, as well as for the indirect products of nitrogen fixation—namely, steaks, hamburgers, and milk.

CONCEPT AND REASONING CHECKS 3

- **a.** Identify the role of microorganisms in the carbon cycle.
- **b.** Identify the role of microorganisms in the sulfur cycle.
- **c.** Identify the role of microorganisms in the nitrogen cycle.

Chapter Challenge C

Cyanobacteria species not only thrive in fresh water lakes and ponds, but are most numerous in the oceans where there can be up to a million cells in just one milliliter of sea water. In the nutrient-poor (oligotrophic) areas, two genera, *Synechococcus* and *Prochlorococcus*, dominate. Their distribution changes according to ocean temperature, light availability, and nutrient status. *Synechococcus* is the hardier of the two and it dominates in water with a latitude of more than 40°; *Prochlorococcus* is the major genus in warmer waters.

QUESTION C:

Because these are oligotrophic areas of the ocean, blooms are not a danger. So, what positive roles could these vast numbers of bacterial cells be playing and how significant is their role to the "daily life" on planet Earth?

Answers can be found on the Student Companion Website in **Appendix D.**

SUMMARY OF KEY CONCEPTS

26.1 Water Pollution Includes Biological Changes Harmful to Water Quality

1. An unpolluted water environment is inhabited by limited numbers of soil bacteria, but a polluted environment contains an enormous variety of heterotrophic organisms from sewage, feces, and industrial sources. **Coliform bacteria**, the gram-negative rods of human and animal intestinal tracts, abound in polluted water. By contrast, a marine environment has halophilic microorganisms as well as psychrophilic and barophilic organisms.
2. Of the three types of pollution, **biological pollution** is of primary interest to the water microbiologist. The **biochemical oxygen demand** (**BOD**) is a measure of the amount of biological pollution. A high BOD indicates the presence of large numbers of organisms needing oxygen for decomposition processes.
3. Concern exists for a variety of bacterial, viral, and protozoal diseases transmitted by water.

26.2 Proper Treatment of Water and Sewage Ensures Safe Drinking Water

4. To prepare water for drinking purposes, municipalities employ various levels of water purification, including **sedimentation**, **filtration**, and **chlorination.** (Fig. 26.6)
5. Sewage also can be treated by different steps according to the needs of the municipalities. **Cesspools** and **septic tanks** are used for local treatment, and variations of **oxidation lagoons**, as well as **secondary** and **tertiary treatments** are used on larger scales. (Fig. 26.7)
6. As much as 99% of microbial activities in nature may occur in **biofilms.** The immobilized populations adhere to many different surfaces. They can be used in such processes as bioremediation and oil spill clean up. Biofilms also can cause disease and be quite resistant to chemicals and antibiotics.
7. To test the effectiveness of purification procedures, several bacteriological tests are available, including the **membrane filter technique**, the **standard plate count**, and the **most probable number** test.

26.3 Microbes Are Indispensable for Recycling Major Chemical Elements

8. In the **carbon cycle**, bacterial and fungal cells are essential to the breakdown of organic matter and the release of carbon back to the atmosphere for recycling. (Fig. 26.11)
9. Anaerobic bacteria are important to the **sulfur cycle** by producing and degrading sulfate. (Fig. 26.12)
10. In the **nitrogen cycle**, many microorganisms release nitrogen from urea, amino acids, and nitrogenous organic matter. Many types of nitrogen-based conversions are performed by microorganisms, and specific bacterial species are essential in the steps of **nitrogen fixation** where nitrogen is brought back into the cycle. Indeed, the processes performed by nitrogen-fixing bacterial species are so essential that life as we know it probably would not exist without this bacterial intervention. (Fig. 26.13)

CHAPTER SELF-TEST

For **STEPS A–D**, answers to even-numbered questions and problems can be found in **Appendix C** on the Student Companion Website at **http://microbiology.jbpub.com/10e**. In addition, the site features eLearning, an online review area that provides quizzes and other tools to help you study for your class. You can also follow useful links for in-depth information, read more MicroFocus stories, or just find out the latest microbiology news.

STEP A: REVIEW OF FACTS AND TERMS

Multiple Choice

Read each question carefully, then select the ***one*** answer that best fits the question or statement.

1. Which one of the following would be considered a coliform?
   A. *Pseudomonas*
   B. *Proteus*
   C. *Enterobacter*
   D. *Streptococcus*
2. Chemical pollution in water would include all the following *except* _____.
   A. pesticides
   B. particulate matter
   C. phosphates
   D. nitrates
3. Intestinal diseases transmitted in water include all the following *except* _____.
   A. cholera
   B. typhoid fever
   C. giardiasis
   D. tetanus
4. During water purification, the jelly-like masses of coagulated material are called _____.
   A. sediments
   B. flocs
   C. schmutzdecke
   D. sludge
5. Raw sewage piped into tanks for waste removal is called _____.
   A. activated sludge
   B. grit
   C. sludge
   D. slime

6. Planktonic bacteria are _____.
   A. individual and independent cells
   B. pathogens in open marine waters
   C. immobilized cells
   D. the cells in a biofilm
7. The most probable number test applies to _____.
   A. detecting bacterial contamination of water
   B. estimating the number of nitrogen-fixing bacteria in soil
   C. the standard number of coliforms allowed in clean water
   D. determining the number of microbes in organic waste
8. Which one of the following is critical to the recycling of carbon?
   A. Photosynthetic microorganisms
   B. Decomposers
   C. Green plants
   D. All the above (**A–C**) are important.
9. Anaerobic microbes important to the sulfur cycle include all the following *except* _____.
   A. *Desulfovibrio*
   B. *Rhizobium*
   C. *Thiobacillus*
   D. *Thiothrix*
10. _____ is a genus that carries out nitrogen fixation, which is the conversion of _______________.
   A. *Rhizobium;* nitrite to nitrate
   B. *Nitrosomonas;* nitrate to nitrite
   C. *Nitrobacter;* nitrate to nitrogen gas
   D. *Rhizobium;* nitrogen gas into ammonia

Using your knowledge of environmental microbiology, consider each characteristic and the three possible choices below. In the space, place the letter or letters of the most appropriate choice(s).

11. Genus (genera) of coliform bacteria _____
   A. *Escherichia*
   B. *Staphylococcus*
   C. *Enterobacter*
12. Where halophilic bacteria live _____
   A. Lake
   B. Ocean
   C. Mountain stream
13. Used as markers for oil drilling _____
   A. Bacteriophages
   B. Radiolarians
   C. Foraminifera
14. Waterborne microbial disease(s) _____
   A. Hepatitis A
   B. Amoebiasis
   C. Hepatitis B
15. Step(s) in water purification _____
   A. Filtration
   B. Chlorination
   C. Sedimentation
16. Found in polluted water _____
   A. *Proteus* species
   B. *E. coli*
   C. HIV
17. Test(s) for oxygen consumption in water _____
   A. SPC
   B. BOD
   C. MPN
18. Type(s) of pollution when microorganisms are present _____
   A. Biological
   B. Physical
   C. Chemical
19. Toxin-producing dinoflagellate(s) _____
   A. *Entamoeba*
   B. *Gambierdiscus*
   C. *Gonyaulax*
20. Produce(s) flocs in water _____
   A. Iron sulfate
   B. Copper sulfate
   C. Aluminum sulfate
21. Needed to perform the standard plate count _____
   A. Culture dishes
   B. Pipettes
   C. Agar medium
22. Possible cause(s) of red tide _____
   A. *Gonyaulax*
   B. *Streptococcus*
   C. *Gymnodinium*
23. Found in anaerobic mud at lake bottom _____
   A. *Giardia*
   B. Diatoms
   C. *Clostridium*
24. Release(s) carbon in carbon cycle _____
   A. Viruses
   B. Fungi
   C. Bacteria
25. Function(s) in nitrogen cycle _____
   A. *Thiobacillus*
   B. *Nitrobacter*
   C. *Thiothrix*

STEP B: CONCEPT REVIEW

26. Evaluate the microbiological differences between unpolluted and polluted water. (**Key Concept 1**)
27. Assess the importance of the **biological oxygen demand** (**BOD**) when measuring polluted water. (**Key Concept 1**)
28. Identify several intestinal diseases and wound infections transmitted by contaminated water. (**Key Concept 1**)
29. Summarize the three sequential steps for **water purification**. (**Key Concept 2**)
30. Discuss what happens in each step of sewage treatment. (**Key Concept 2**)
31. Describe **biofilms** and how these organisms differ from **planktonic** forms. (**Key Concept 2**)
32. Identify three ways that water can be tested for **indicator organism** contamination. (**Key Concept 2**)
33. Summarize the **carbon cycle** and identify the role of microorganisms in the cycle. (**Key Concept 3**)
34. Describe how microorganisms affect the **sulfur cycle**. (**Key Concept 3**)
35. Assess the critical role of microorganisms to the **nitrogen cycle**. (**Key Concept 3**)

STEP C: APPLICATIONS AND PROBLEMS

36. In August 1989, two men were tragically killed while working in an industrial septic tank during cleaning. The cause of death was listed as methane poisoning. What was the source of the methane, and how was it related to the septic tank?

37. In 1978, Legionnaires' disease broke out in the garment district of New York City and was given front-page treatment in the press. During that same period, the coliform count rose dramatically in water in the Murray Hill section of the city, but this was given minor coverage in the papers. Many public health officials believed that the Murray Hill problem posed the more substantial threat. Do you agree? Why?

38. You decide to build a new home in the country and you have the choice of installing a cesspool or a septic tank. Which might you be inclined to choose? Why?

39. Working for the city parks and recreation department, you have two swimming pools to inspect: an Olympic-sized pool for swimmers, and a small wading pool for toddlers. In which pool does the greater potential for disease transmission exist, and what precautions should you take to limit the transmission of microorganisms?

STEP D: QUESTIONS FOR THOUGHT AND DISCUSSION

40. When sewers were constructed in New York City in the early 1900s, engineers decided to join storm sewers carrying water from the streets together with sanitary sewers bringing waste from the homes. The result was one gigantic sewer system. In retrospect, was this a good idea? Why?

41. The victims of disease, both animals and people, are buried underground, yet the soil is generally free of pathogenic organisms. Why?

42. What information might you offer to dispute the following four adages common among campers and hikers? (1) Water in streams is safe to drink if there are no humans or large animals upstream. (2) Melted ice and snow is safer than running water. (3) Water gurgling directly out of the ground or running out from behind rocks is safe to drink. (4) Rapidly moving water is germ free.

43. Some years ago, the syndicated columnist Erma Bombeck wrote a humorous book entitled "The Grass Is Always Greener Over the Septic Tank." (The title was an adaptation of the expression "The grass is always greener on the other side of the fence.") Indeed, the grass is often greener over the septic tank. Why is this so?

44. The author of a biology textbook writes: "Because the microorganisms are not observed as easily as the plants and animals, we tend to forget about them, or to think only of the harmful ones . . . and thus overlook the others, many of which are indispensable to our continued existence." How do the carbon, sulfur, and nitrogen cycles support this outlook?

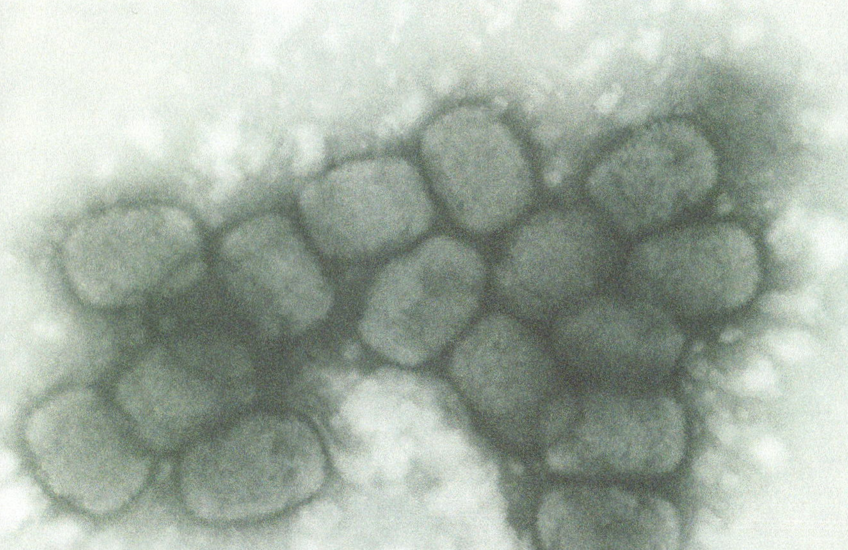

Industrial Microbiology and Biotechnology

27

CHAPTER PREVIEW

Never underestimate the power of the microbe.
—Microbiologist Jackson W. Foster of the University of Texas

Humans probably discovered alcoholic beverages by accident. It is conceivable that sunlight warmed some sort of fallen grape or other fruit and accelerated the fermentation of its juices by yeasts. Humans must have sampled this "spoiled" fruit with curiosity, and if the taste of the aromatic concoction was not especially pleasing, the euphoric feeling that followed probably brought them back for more. By trial and error, humans discovered the important factors in fermentation and soon learned to control the process. In doing so, they became the first industrial microbiologists.

How fermentation occurs was worked out by Louis Pasteur when he studied the souring of French wines. With this understanding, the process of industrial fermentation, using the metabolic products of microorganisms for commercial purposes was developed. By the 1930s, several types of fermentations were being carried out to produce ethyl alcohol and butyl alcohol. For the British war effort, Chaim Weizmann's development of a microbial fermentation process for acetone using *Clostridium acetobutylicum* was critical toward the manufacture of gunpowder.

However, industrial microbiology is not restricted to alcohol fermentations. It includes bread baking and cheese production, as well as many of the foods resulting from fermentation. Industrial fermentation enhances the metabolic reactions already present in (or genetically engineered into) microbes. **Biocatalysis** is the term used to describe the metabolic reactions carried out at this industrial scale.

To produce a significantly useful amount of product from microorganisms, tremendously large numbers of microorganisms are needed. This means scaling up microbial metabolic processes to a level significant enough to produce

Image courtesy of Dr. Fred Murphy/CDC.

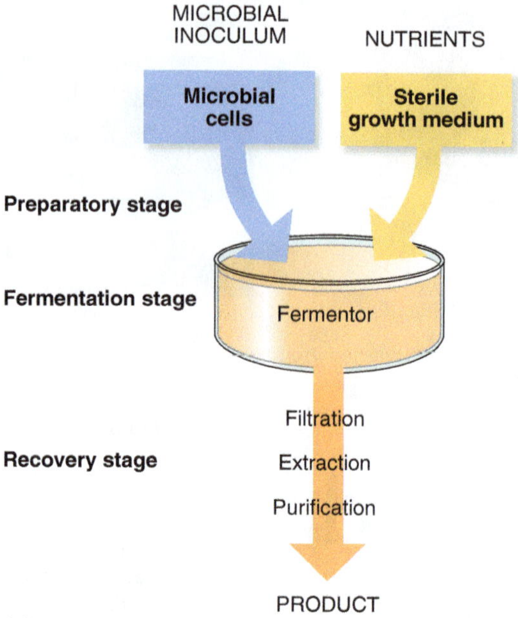

(A)

(B)

FIGURE 27.1 The Industrial Fermentation Process. (**A**) A general industrial process scheme for biocatalysis and product recovery. (**B**) A pharmaceutical technician monitors a series of fermentors to ensure a maximum yield of microbial product. © Maximilian Stock LTD/Phototake/Alamy Images. »» What types of conditions must this technician monitor?

the valuable commercial product (FIGURE 27.1A). On this large scale, the term *fermentation* has come to have a slightly different definition. **Industrial fermentation** procedures are not necessarily anaerobic processes; rather, they are large-scale microbial methods that may be aerobic or anaerobic.

Such biocatalytic processes are carried out in large containers or tanks called **bioreactors** or **fermentors** (FIGURE 27.1B). Technicians or scientists must continually monitor the growth conditions in these containers to ensure that microbes continue to either grow or reach a state in which maximal product is produced. Temperature, pH, oxygen, and nutrients—all of the typical physical and chemical conditions that apply to growing microbes in a culture tube—apply here, just on a grander scale.

The exploitation of microbes for use in industrial processes is closely linked to **biotechnology**. Thus, many of the products of industrial microbiology contribute to public health, interrupt the spread of disease, or improve the quality of life. Microbial cells or their products are used, for example, to produce enzymes, insecticides, food additives, protein hormones, antibiotics, monoclonal antibodies, interferon, and the myriad products of genetic engineering that are at the heart of the biotechnology industry today.

As we see in this chapter, industrial microbiology is an extremely diversified field, in which inexpensive raw materials are converted to valuable commodities through the metabolism of microorganisms.

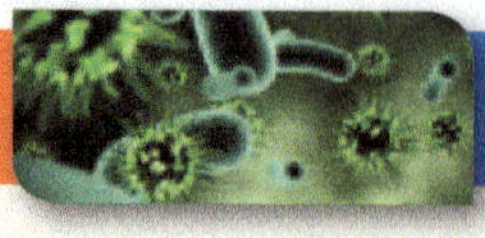

Chapter Challenge

Many of the foods we eat are the result of microbial metabolism, especially fermentation. However, there are many other products and medicines that we use every day that are the result of microbial activity. Make a short list right now of those products and medicines that you can think of that you have used. As we progress through this chapter, check off the products on your list that are described and add to your list as we discover other products that you forgot about or did not know had a microbial origin.

KEY CONCEPT 27.1 Microorganisms Are Used to Produce Many Industrial Products

Certain properties make microorganisms well suited for industrial processes. Microorganisms not only possess a broad variety of enzymes to make an array of chemical conversions possible, they also have a relatively high metabolic activity that allows conversions to take place rapidly. In addition, they have a large surface area for the quick absorption of nutrients and release of end products. Moreover, they usually multiply at a high rate, as evidenced by the 20-minute generation time for *Escherichia coli* under ideal conditions.

In the industrial process, microorganisms act like chemical factories. To be effective, they should liberate a large amount of a single product that can be extracted and purified efficiently. The organisms should be easy to maintain and cultivate, and should have genetic stability with infrequent mutations. Their value is enhanced if they can grow on an inexpensive, readily available medium that is a byproduct of other industrial processes. For example, a large amount of whey is produced in cheese manufacturing, and microorganisms that convert whey components to lactic acid add to the overall profit of the cheese industry. FIGURE 27.2 displays some possible conversions from a single metabolic substance.

Microorganisms Produce Many Useful Organic Compounds

Microorganisms are used in industry to produce a variety of organic compounds, including acids, growth stimulants, and enzymes. In some cases, the production results from a microorganism manufacturing many thousands of times the amount of product necessary for its own metabolism.

Microorganisms may produce one of two types of **metabolites** that are important to industrial microbiology. **Primary metabolites** are directly involved in the normal growth, development, and reproduction of the microbe. Pyruvate and the end products of the fermentation pathways shown in Figure 27.2 are examples of primary metabolites.

Metabolites: Small molecules involved in, or byproducts of, metabolism.

Secondary metabolites form the bulk of products of industrial interest. These metabolites

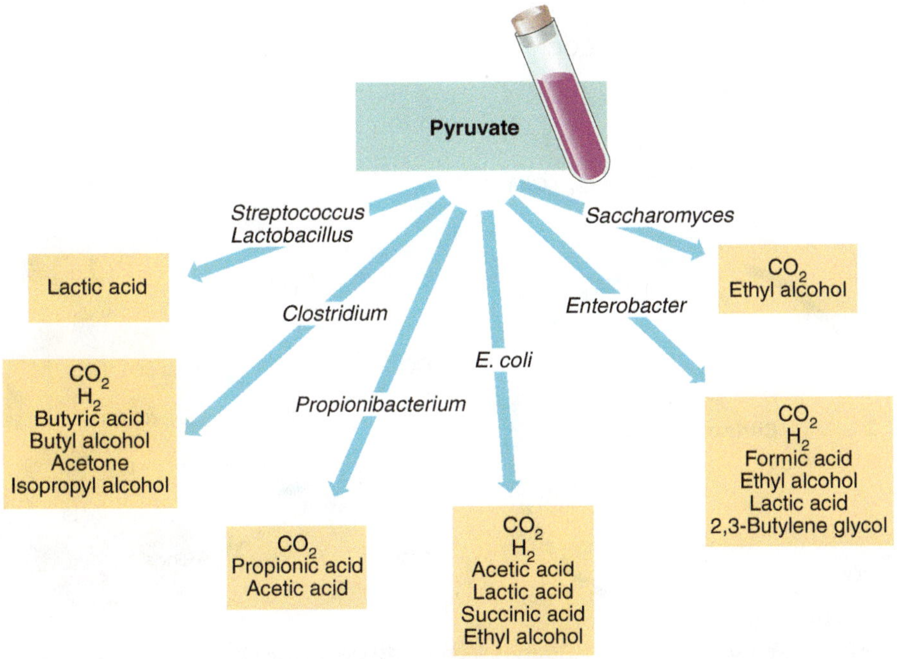

FIGURE 27.2 **The Products of Fermentation.** Industrial microbiology includes the metabolic conversions of organic compounds to other organic compounds by microorganisms. These conversions are called fermentations, which in this case are taking place in the absence of oxygen gas. Pyruvate is the starting point for the conversions shown here. »» What metabolic pathway would contribute to the production of pyruvate?

are not directly involved in growth and reproduction, but usually have important ecological functions and often are species specific. Secondary metabolites usually are produced near or at the end of the microbe's growth phase; that is, in the stationary phase.

One of the first secondary metabolites to be made in bulk by industrial fermentation was citric acid. Manufacturers use this organic compound in soft drinks, candies, inks, engraving materials, and a variety of pharmaceuticals, such as anticoagulants and effervescent tablets (e.g., Alka-Seltzer®). The organism most widely used for citric acid production is the fungus *Aspergillus niger*. Microbiologists inoculate the mold into a medium of cornmeal, molasses, salts, and inorganic nitrogen in huge aerobic fermentors. The absence of a citric acid cycle enzyme in the fungus prevents the metabolism of citric acid (citrate) into the next component of the cycle, and the citric acid accumulates in the medium. FIGURE 27.3 outlines this chemistry.

Another important microbial product is lactic acid, a compound employed to preserve foods, finish fabrics, prepare hides for leather, and dissolve lacquers. Lactic acid is commonly produced by bacterial activity on the whey portion of milk. *Lactobacillus bulgaricus* is widely used in the fermentation because it produces only lactic acid from lactose.

Gluconic acid (gluconate), another valuable organic acid, is useful in medicine as a carrier for calcium, because gluconic acid is easily metabolized in the body, leaving a store of calcium for distri-

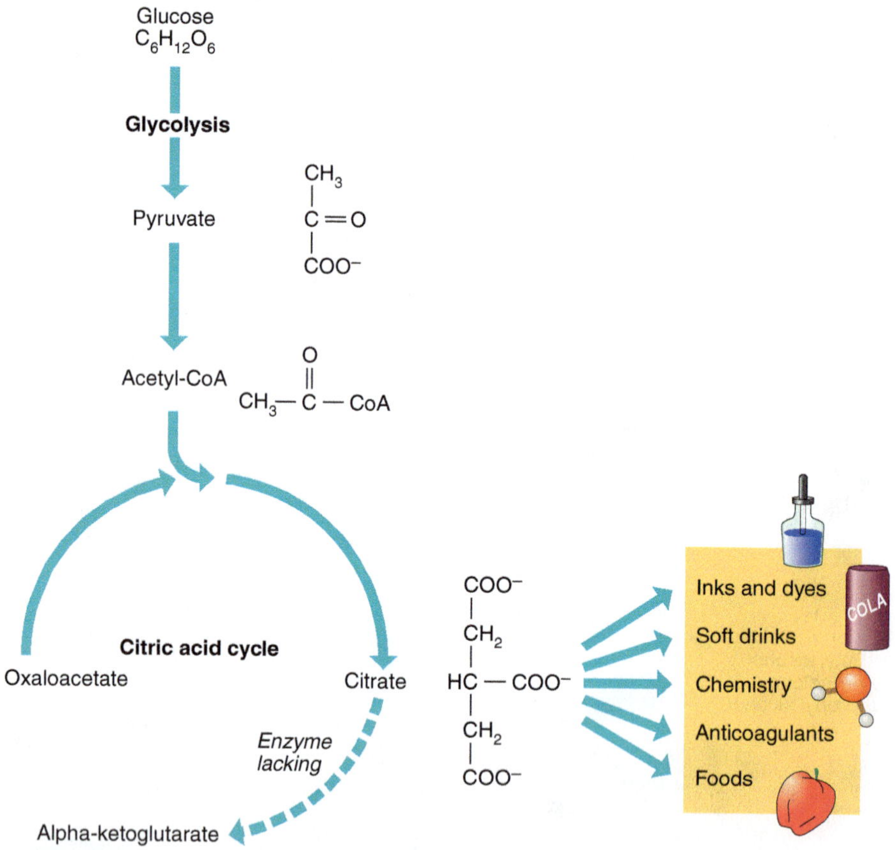

FIGURE 27.3 **The Chemistry of Citrate Production.** *Aspergillus niger* is grown in a mixture of nutrients, where it digests glucose into pyruvate. The pyruvate then is converted to acetyl-CoA, which condenses with oxaloacetate in the citric acid cycle to yield citric acid or citrate. However, the chemistry goes no further, because the next enzyme in the cycle is absent. Citric acid therefore accumulates and is isolated for use in various products, as shown. »» Because the rest of the citric acid cycle does not occur in this strain of *Aspergillus*, how does the organism make ATP?

bution. This acid is produced from carbohydrates by *A. niger* and by species of the bacterial genus *Gluconobacter* cultivated in fermentors. Calcium gluconate also is added to the feed of laying hens to provide calcium to strengthen the eggshells.

When the amount of amino acid produced by a microorganism exceeds what the microbe needs, the remainder is excreted into the environment. Such is the case with glutamic acid (glutamate) produced by certain species of *Micrococcus*, *Arthrobacter*, and *Brevibacterium*. Glutamic acid is a valuable food supplement for humans and animals, and its sodium salt, monosodium glutamate, is used in food preparations.

In the industrial production of lysine, another amino acid, two organisms are involved. *E. coli* is first cultivated in a medium of glycerol, corn steep liquor, and other ingredients, resulting in the accumulation of diaminopimelic acid (DAP). Several days later, *Enterobacter aerogenes* is added to the fermentor. This organism produces an enzyme that removes the carboxyl group from DAP to produce the lysine used in breads, breakfast cereals, and other foods.

Two important vitamins, riboflavin (vitamin B_2) and cyanocobalamin (vitamin B_{12}), also are products of microbial growth. Riboflavin is a product of *Ashbya gossypii*, a mold able to produce 20,000 times the amount of riboflavin it needs for metabolism. Cyanocobalamin is produced by selected species of *Pseudomonas*, *Propionibacterium*, and *Streptomyces* grown in a cobalt-supplemented medium. The vitamin prevents pernicious anemia in humans and is used in bread, flour, cereal products, and animal feeds.

Microorganisms Also Produce Important Enzymes and Other Products

The production of microbial enzymes for commercial exploitation has been an important industry since the emergence of industrial microbiology. Currently, over two dozen types of microbial enzymes are in use, and several others are in the research or developmental stage. Industrial enzymes have reached an annual market of $1.6 billion.

Among the important microbial enzymes used commercially are amylase, pectinase, and several proteases (TABLE 27.1). Amylase is produced by the mold *Aspergillus oryzae*. It is used as a spot remover in laundry presoaks, as an adhesive, and in baking, where it digests starch to glucose. Pectinase, a product of a *Clostridium* species, is employed to ret flax for linen. In this process, manufacturers mix the flax plant with pectinase to decompose the pectin "cement" holding cellulose fibers together. The cellulose fibers then are spun into linen. MicroFocus 27.1 describes a more traditional process for retting. Pectinase also is used to clarify fruit juices.

Corn steep liquor: A pure extract from corn that contains high concentrations of free amino acids.

TABLE

27.1 Some Enzymes and Their Microbial Sources Used in Commercial Markets

Commercial Market	Enzyme	Source	Use
Dairy	Proteases Lactase	Bacteria and molds Molds and yeasts	Hydrolysis of whey proteins; coagulant in cheese production Hydrolysis of lactose to produce lactose-free milk
Brewing	Cellulases Amylases and proteases	Molds Bacteria and molds	Liquefaction and clarification processes
Wine and juices	Pectinases	Bacteria and molds	Increase juice yield and clarification
Meat	Proteases	Bacteria and molds	Meat tenderizing
Confectionary	Invertase	Yeasts	Liquefaction of sucrose
Textiles	Pectinase	Bacteria and molds	Flax retting
Pulp and paper	Xylanases, hemicellulases, lipases	Molds and yeasts	Cleaner and more efficient pulp and paper processing
Detergents	Proteases, amylases, lipases, cellulases	Bacteria and molds	Better cleansing of laundry and dishes

MICROFOCUS 27.1: History
It Smelled Bad, but It Worked

In past centuries, industrial pectinase was not available for retting flax, nor was protease available for bating hides. Nevertheless, the processes were carried on efficiently and successfully.

The retting process began by bundling flax plants and drying them in stacks. The stacks were then placed in a long trench several feet deep, covered with water, and weighted down with stones to exclude as much air as possible. After a few days, the water turned black, and an unmistakable stench signaled that retting was taking place. Two weeks later the flax was so soft and pliable that the fibers could be removed easily by pounding with wooden blocks. Today's microbiologists point out that *Clostridium* species were probably producing pectinase in those trenches.

The method for treating hides was equally messy. Skins were mixed with dog or fowl manure and set aside to cure. Fragments of tissue and hair gradually dissolved in the muck, and soon the hide became soft and pliable. Apparently, the proteases from fecal bacteria were responsible for the digestion. Bating hides was another smelly process, to be sure, but like the method for retting, it was usually reliable.

Proteases are a group of protein-digesting enzymes produced by *Bacillus subtilis, A. oryzae*, and other microorganisms. Certain proteases are used for bating hides in leather manufacturing, a process in which organic tissue is removed from the skin to yield a finer texture and grain. Other proteases find value as liquid glues, laundry presoaks, meat tenderizers, drain openers, and spot removers.

One of the most appreciated but lesser known uses of a microbial enzyme is in making soft-centered chocolates. Invertase, an enzyme from yeast, is mixed with flavoring agents and solid sucrose, and then covered with chocolate. The enzyme converts some of the sucrose to liquid glucose and fructose, forming the soft center of the chocolate.

In medical microbiology, doctors use another microbial enzyme, **streptokinase**, to break down blood clots formed during a heart attack. Still another enzyme, **hyaluronidase**, is used to facilitate the absorption of fluids injected under the skin. The microbial roles for these enzymes as virulence factors are discussed in another chapter.

Although most natural food flavoring ingredients are produced by traditional processes using plants, new biotechnology methods have made it possible to produce novel flavoring ingredients by converting relatively cheap starting materials into higher-value flavor and aroma additives. The latter are used in foods, beverages, cosmetics, and other consumer items; examples include fruit, peach, and coconut flavoring agents called **lactones**. Although lactones can be generated from long-chain fatty acids derived from sweet potatoes, the fatty acids occur in too limited a quantity and are expensive to modify. To circumvent this problem, microbiologists use species of *Mucor* and other fungi to convert medium-chain fatty acids to compounds other microorganisms can easily transform to lactones. Another example is the methylketones, which confer strong cheese-associated flavors in dairy products. The flavoring agents are derived industrially from *Penicillium roqueforti* incubated in lipase-treated milk fats.

In addition to the major products we have surveyed, microorganisms provide a number of specialized materials. Typical of the miscellaneous microbial products is **alginate**, a sticky substance used as a thickener in ice cream, soups, and other foods. Other products of microbial origin are some perfumes. Musk oil, for example, is prepared from ustilagic acid, a product of the mold *Ustilago zeae*, which, ironically, causes smut disease (see Chapter 11). Moreover, there are numerous pharmaceutical products derived from the ergot poisons of the mold *Claviceps purpurea*. These derivatives are prescribed to induce labor, treat menstrual disorders, and control migraine headaches.

CONCEPT AND REASONING CHECKS 1

a. Assess the importance of microbial secondary metabolites to industry and commercial markets.

b. Identify some of the enzymes for industry that are produced by microorganisms.

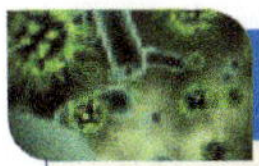

Chapter Challenge A

Microorganisms produce many metabolic products that have use directly or indirectly with human activities. These included both primary and secondary metabolites.

QUESTION A:

Add to your list those primary and secondary metabolites with which you have had an "encounter"; that is, used directly or indirectly through its addition to another product. Your list might get a bit long!!

Answers can be found on the Student Companion Website in **Appendix D.**

KEY CONCEPT 27.2 Alcoholic Beverages Are Products of Fermentation

The fermentation of beer, wine, and other alcoholic beverages is one of the most venerable and universal of human domestic activities. The origin of beer fermentation, for example, has been traced as far back as 4000 BC, when legend tells us that Osiris, the god of agriculture, taught Egyptians the art of brewing once they had learned how to farm the land (MICROFOCUS 27.2). Wine production apparently has an equally long history because archaeologists have discovered evidence of grape cultivation in the Nile Valley during the same period. Furthermore, scientists have found evidence of wine in jars excavated from an Iranian site 7,000 years old. Thus, it is conceivable that the Egyptians, Sumerians, Assyrians, and other Near East peoples were among the earliest consumers, if not connoisseurs, of alcoholic beverages.

Beer Is Produced by the Fermentation of Malted Barley

As early as 3400 BC, a tax was placed on beer in the ancient Egyptian city of Memphis on the Nile River. The Greeks later brought the art of brewing to Western Europe, and the Romans refined it. Indeed, the main drink of Caesar's legions was beer. During the Middle Ages, monasteries were the centers of brewing, and by the 1200s, breweries and taverns were commonplace in Great Britain.

The word *beer* is derived from the Anglo-Saxon *baere*, meaning "barley." Thus, beer is traditionally a product of yeast fermentations of barley grains. However, yeasts cannot digest barley starch, and, therefore, it must be predigested for them (FIGURE 27.4). This is accomplished in the process of **malting**, where barley grains are **steeped** in water while naturally occurring enzymes digest the starch to simpler carbohydrates, principally maltose (malt sugar).

Steeped: Immersed in liquid to soften or extract something.

The malt is ground with water to achieve further digestion of the starch. This process, called **mashing**, often includes corn as a starch supplement. Brewers then remove the liquid portion, or **wort**, and boil it to inactivate the enzymes. Dried petals of the vine *Humulus lupulus*, commonly called **hops**, are added to the wort, giving it flavor, color, and stability. Hops also prevents contamination of the wort, because the leaves contain at least two antimicrobial substances. At this point, the fluid is filtered, and yeast is added in large quantities.

The yeast usually employed in beer fermentation is one of two species of *Saccharomyces* developed for centuries by brewers. One species, *S. cerevisiae*, gives a uniform dark cloudiness to beer and is carried to the top of the fermentation vat by foaming carbon dioxide. This yeast, therefore, is called a "top yeast." It is used primarily in English-type brews such as ale and stout. The second species, *S. carlsbergensis*, ferments the malt more slowly and produces a lighter, clearer beer, having less alcohol. This yeast sediments and is thus called a "bottom yeast." Its product is pilsner or lager beer. Almost three-quarters of the world's beer is lager beer.

A normal fermentation requires approximately 7 days in a fermentation tank. The young beer

MICROFOCUS 27.2: History
This hqt's for You!

Beer, called hqt by the ancient Egyptians, was a very important drink. They often used beer in religious ceremonies and, because water could be a source of illness, they served beer at mealtimes to both adults and children. In fact, archaeologists have discovered that workmen at the great pyramids had five types of beer that they drank three times a day. It was the staple drink of the poor (wages often were paid in beer), a drink of the pharaohs, and a drink offered to the gods.

Because of the prevalence of beer in Egyptian life, many Egyptologists have studied beer residue from ancient Egyptian vessels. The traditional view held that their beer was made by crumbling lightly baked well-leavened bread into water. After straining through a vat, the water was allowed to ferment because of the yeast from the bread. It then was flavored with date juice or honey.

In 1996, these traditional views were challenged. Delwen Samuel, an archaeobotanist at the University of Cambridge, examined just what cereals and grains the Egyptians used to brew beer by looking at 2,000-year-old beer residues using a scanning electron microscope. Her findings suggested that the ancient Egyptians used barley to make malt and a type of wheat, called emmer, instead of hops. She says they heated the mixture and then added yeast and uncooked malt to the cooked malt. After adding the second batch of malt, the mixture was allowed to ferment. In her analysis, Samuel says she found no traces of flavorings.

Samuel and her colleagues tried brewing the beer using the recipe derived by the analysis. They brewed it at a modern brewery and found the beer to be fruity and sweet because it lacked the bitterness of hops. The beer was reported to have an alcoholic content of between 5% and 6%. Samuel gave her recipe to Scottish and Newcastle Breweries in London that then bottled a limited edition, 1,000-bottle batch of "King Tut Ale." It was sold at Harrods department store for $100 per bottle, the proceeds going toward further research into Egyptian beer making.

In 2002, a major Japanese brewery claimed to have recreated a 4,000-year-old Egyptian beer by following a recipe from ancient hieroglyphs in Egyptian tomb paintings. Kirin Brewery Company Ltd., Japan's second-largest beer maker, said the 30 liters of brewed beer was dark brown, contained no froth, had a strong sour taste, and an alcohol content of about 10%. The company does not plan to sell the beer commercially—it was developed for research purposes only!

then is transferred to vats for secondary aging, or **lagering**, which may take an additional 6 months. If the beer is intended for canning or bottling, it is pasteurized at 60°C for 55 minutes to kill the yeasts, or filtered through a membrane filter. Some yeast is used to seed new wort, and the remainder may be dried for animal feed or pressed to tablets for human consumption. The alcoholic content of beer is approximately 4%.

Wine Is Produced by the Fermentation of Fruit or Plant Extracts

During the Middle Ages and the centuries thereafter, wine was called *aqua vitae*, the "water of life." The title was appropriate because wine was one of the few safe things to drink. Indeed, until the late 1800s, safe drinking water was virtually nonexistent in the Western world (note that the Bible makes no references to water for drinking purposes). Wine, by contrast, was generally free of pathogens due to its acidity and alcohol content, and it provided a few minerals and vitamins to the diet, while serving as a pain reliever. Of course, it also bred a society somewhat inebriated much of the time.

Essentially, all wines are derived from the natural conversion of grape or other fruit sugars to ethyl alcohol by fermentation with *Saccharomyces*. Wild yeasts, which occur naturally on the grapes, can produce undesirable qualities in a wine, so they are killed by adding sulfur dioxide ("sulfites") during the production process.

Wine may be made from fruit, fruit juice, or plant extracts such as dandelions. Among the grapes, the species *Vitis vinifera* is recognized as the highest-quality fruit. The winemaking process begins with crushing to produce the juice, or **must** (FIGURE 27.5). For red wine, black grapes are used, including skins and sometimes, the stems. White wine, by contrast, is made from black or white

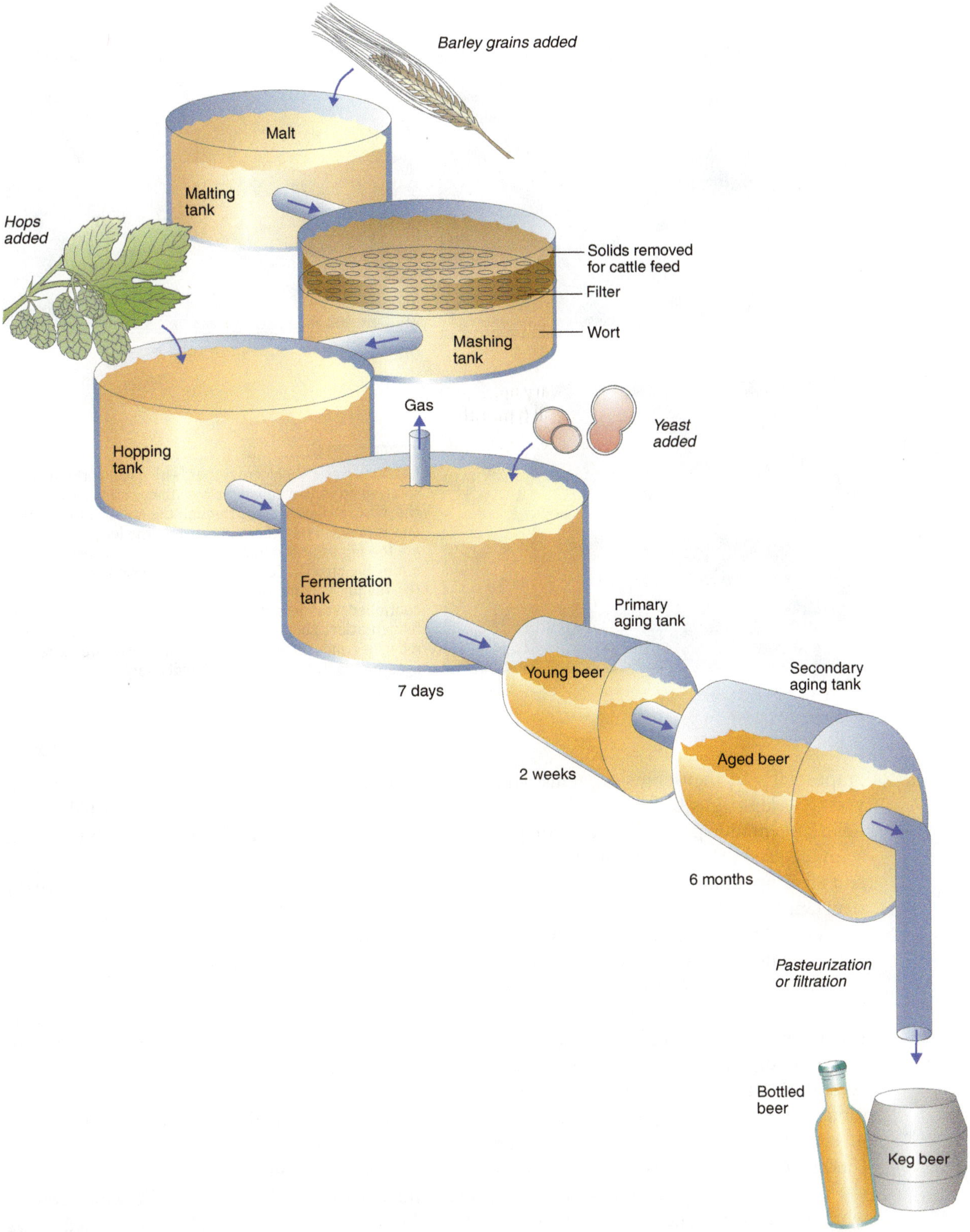

FIGURE 27.4 **A Generalized Process for Producing Beer.** Barley grains are held in malting tanks while the seeds germinate to yield fermentable sugars. The digested grain, or malt, is then mashed in a mashing tank and the fluid portion, the wort, is removed. Hops are added to the wort in the next step, followed by yeast growth and alcohol production during fermentation. The young beer is aged in primary and secondary aging tanks. When it is ready for consumption, it is transferred to kegs, bottles, or cans. »» Why must beer be pasteurized or filtered before packaging?

FIGURE 27.5 The Large-Scale Production of Wine. (**A**) A view of a stemmer-crusher, which is a perforated, rotating drum that separates the stems and leaves from the must. The must drops through the holes in the drum and is pumped into the winery, while the leaves and stems are too big to fit through the holes so they continue on out of the crusher. Courtesy of Jeffrey Pommerville. (**B**) The juice, skins, and seeds go into large, temperature-controlled, stainless steel fermentation tanks for primary fermentation. Courtesy of Jeffrey Pommerville. (**C**) Many red wines and white wines, like chardonnay, are barrel-aged for months or years. © LiquidLibrary. »» Why are wooden barrels used for aging wines?

grapes without their skins or stems. Cultured yeasts (*S. ellipsoideus*) are added to begin the alcoholic fermentation process. Anaerobic conditions soon are established as carbon dioxide evolves and takes up all the air space within the vats or steel tanks.

Alcohol production requires only a few days, but the aging process in wooden casks may go on for weeks or months. During this time wine develops its unique flavor, aroma, and bouquet. These result from the array of alcohols, acids, aldehydes, and other organic compounds produced by the yeast during aging. In fact, wine contains thousands of compounds, a few of which have been identified (MicroFocus 27.3). Soil and climate conditions (the *terroir*, as it is called in French) also contribute to the wine because they determine what organic compounds are present in the grape. The type of yeast and nature of wood derivatives from the fermentation casks are other determining factors. Thus, there are "vintage years" and poor years. For mass production, wine is pasteurized to increase its shelf life, filtered, and bottled.

The broad variety of available wines results from modifications of the basic fermentation process. In dry wines, for example, most or all of the sugar is metabolized, while in sweet wines, fermentation is stopped while there is residual sugar. Sparkling wines, including champagne, sparkle because of a second fermentation taking place inside the bottle. For a sweet sauterne, vintners enhance the sugar content of grapes by a controlled infection with the mold *Botrytis cinerea*. The mold literally sucks water out of the grapes, thereby increasing the sugar concentration.

The strongest natural wines measure about 15% alcohol because yeasts cannot tolerate alcohol much above this level. Most table wines average about 10% to 12% alcohol, with fortified wines reaching 22% alcohol. In fortified wines, brandy or other spirits are added to produce such wines as port, sherry, and Madeira.

MICROFOCUS 27.3: Being Skeptical
The Benefits of Red Wine

In 1992, scientists from the Bordeaux region of France did a population and epidemiological study and noted that while French and other Mediterranean peoples ate large amounts of fatty foods, they suffered a relatively low incidence of coronary artery disease. A 1996 report by the television news magazine program *60 Minutes* also pointed out that fatty meats, creams, butters, and sauces had little apparent effect on French hearts. This so-called "French Paradox" was due, in part, to the apparent medicinal properties of wine, especially drinking red wine. So, is red wine good for your health?

In 1996, researchers identified phenol-based compounds in red wine that inhibited the oxidation of low-density lipoproteins (LDLs) in the blood, and by doing so, prevented the buildup of cholesterol and blood platelets in the arteries. The scientists pointed out that red wine contains more phenolics than white wine, and far more than beer. The phenolics also were present in other foods (e.g., raisins and onions), but not in the quantity found in the skins of grapes used for red wine, which concentrate even more after fermentation has taken place.

A compound known as resveratrol, which belongs to the family of phenols, naturally exists in grapes and red wine. This compound was shown to extend the lifespan of yeast cells by up to 80% and might help explain why moderate consumption of red wine has been linked to a lower incidence of heart disease in humans and cancer in experimental mice.

Red wine and the resveratrol it contains also has an inhibitory effect on the foodborne pathogens *Escherichia coli, Salmonella enterica,* and *Listeria monocytogenes. Helicobacter pylori,* which is the main cause of stomach ulcers, is especially susceptible.

However, red wine is not for everyone. Indeed, alcohol should be avoided by pregnant women, people taking medications, those under the legal drinking age, and anyone with a family history of alcoholism. For these and for anyone else wishing to stay away from alcohol, alcohol-free red wines are appearing in the marketplace. They offer the opportunity to take advantage of a natural health ingredient while enjoying a glass of nature's bounty. Perhaps Louis Pasteur was correct when he said, "Wine is the most healthful and hygienic of beverages."

Distilled Spirits Contain More Alcohol than Beer or Wine

Distilled spirits are alcoholic beverages containing ethyl alcohol resulting from distillation of a fermented substance, such as fruits or grains. Gin, vodka, tequila, and whiskey are a few types of spirits. Such spirits contain considerably more alcohol than beer or wine. Most of the world measures the alcohol content by its percentage of the total liquid; that is, by volume. However, in the United States, distilled beverages are designated with a "proof number," which is twice the percentage of the alcohol content. For example, a 90 proof product contains 45% alcohol.

The production of distilled spirits begins like a wine fermentation. A raw product is fermented by *Saccharomyces* species, then aged, and finally matured in casks. At this point, the process diverges as manufacturers concentrate the alcohol by a distillation apparatus using heat and vacuum pressure. Next, they mature the product in wooden casks to introduce unique flavors from various chemicals, such as aldehydes and volatile acids. Finally, the alcohol is standardized by diluting it with water before bottling.

Four basic types of distilled spirits are produced: brandy, whiskey, rum, and neutral spirits. Brandy is made from fruit or fruit juice, while rum is produced from molasses. Whiskey is a product of various malted cereal grains, such as scotch from barley, rye from rye grain, and bourbon from corn. The final type, neutral spirits, includes vodka, made from potato starch and left unflavored; gin, flavored with the oils of juniper berries; and tequila, containing the fermented juice of the blue agave.

CONCEPT AND REASONING CHECKS 2

- **a.** How does a lager differ from an ale, microbiologically speaking?
- **b.** Explain why the alcohol produced in wine represents a primary metabolite.
- **c.** Summarize how distilled spirit production produces a higher alcohol content than beer or wine.

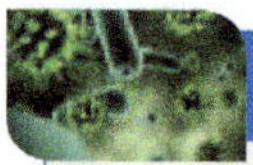

Chapter Challenge B

You may or may not be a wine or beer drinker, or partake of distilled spirits. Yet all are the product of microbial action and alcohol fermentation. So, if you wish you can add these to your list if you haven't already.

QUESTION B:

Try to wrap your head around the reality that, partaking in alcoholic beverages or not, these fermentation products (and distilled products) of microbial origin have a large influence on human behavior, both financially and societally. List some of the impacts they have daily.

Answers can be found on the Student Companion Website in **Appendix D.**

KEY CONCEPT 27.3 Microorganisms Also Produce Many Other Valuable Commercial Products

In addition to the products we have discussed, microorganisms are the sources of antibiotics and a number of valuable insecticides. Moreover, they are the producers of enzymes that break down natural and synthetic wastes in bioremediation, and they are the biological factories for the genetic engineering technology revolutionizing the biotechnology industry today. In the final section of this text, we study the methods for antibiotic and insecticide production and bioremediation, while discussing some details of the genetic engineering process.

Many Antibiotics Are the Result of Industrial Production

Penicillin was the first antibiotic to be produced on an industrial scale. In 1941, Robert H. Coghill of the Fermentation Division of the United States Department of Agriculture (USDA) suggested adapting the deep-tank method used to produce vitamins to produce penicillin. In the ensuing months, he offered several modifications to stimulate the growth of *Penicillium notatum* and increase the penicillin yield. For example, corn steep liquor in the culture medium increased the output 20 times, and the substitution of lactose for glucose made penicillin production still more efficient. Moreover, the search for a higher-yielding producer of the drug led researchers to *Penicillium chrysogenum*, a mold isolated from a rotten cantaloupe purchased in a Peoria, Illinois, supermarket. Treatment with ultraviolet light resulted in a mutant with still higher penicillin yields. By 1943, the United States was producing enough penicillin for the Allied forces, and by 1945, sufficient amounts were available for the civilian population.

To the present time, over 8,000 antibiotic substances have been described and approximately 100 such drugs are available to the medical practitioner. Although most antibiotics are produced by species of *Streptomyces* (FIGURE 27.6), several are also products of *Penicillium* or *Bacillus* species. The worldwide annual production of antibiotics exceeds 500 metric tons, half of which is prescribed for human use.

Antibiotic production may involve fermentation processes producing natural antibiotics or **semisynthetic drugs**, which represent a modification to a natural antibiotic. This is especially true for the penicillins (see text on antimicrobial drugs). The fermentation process is carried out in large aerated, stainless steel fermentors. A typical fermentor may hold 110,000 liters of medium. Older methods employed enormous mats of fungi or actinomycetes on the surface of the tank. Newer technology, however, employs small fragments of submerged fungal hyphae or bacterial cells, rotated and agitated in the medium with a constant stream of oxygen. After several weeks of growth, the stationary phase microorganisms produce the antibiotic. At the appropriate time, the antibiotic is filtered out, extracted, and purified from the growth medium. The remaining brown mash of microorganisms may be dried and sold

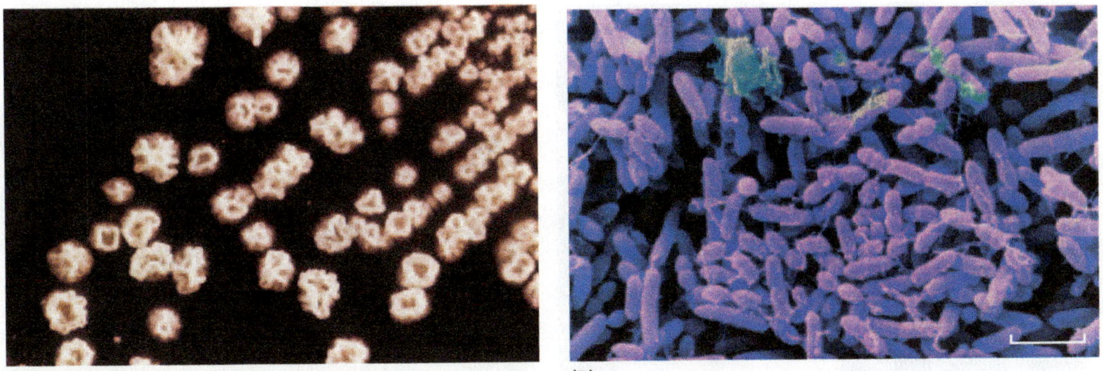

(A) (B)

FIGURE 27.6 **An Antibiotic Producer.** Many modern antibiotics are produced by species of the soilborne *Streptomyces*. (**A**) Colonies of *S. griseus*, the organism from which Selman Waksman isolated streptomycin in the 1940s. © Christine Case/Visuals Unlimited. (**B**) A false-color scanning electron micrograph of *S. griseus*, showing the chains of cells. (Bar = 10 μm.) © Scimat/Photo Researchers, Inc. »» Why is *Streptomyces* such a prolific antibiotic producer?

as an animal feed additive. Another alternative is to process it for use as human food.

Some Microbial Products Can Be Used to Control Insects

To be useful as an insecticide, a toxin from a microorganism should be relatively specific for an insect pest and should act rapidly. It should be stable in the environment and easily dispensed as well as inexpensive to produce. It helps if its odor is pleasant.

In the early part of this century, a scientist named G. S. Berliner discovered that sporulating cells of a *Bacillus* species were inhibitory to moth larvae. Berliner named the organism *Bacillus thuringiensis* after the German state Thuringia where he lived. The *Bacillus* remained in relative obscurity until recently, when scientists learned that older cells of *B. thuringiensis* produce toxic crystals during the process of sporulation (FIGURE 27.7A). The toxic substance, an alkaline protein, is deposited on leaves and ingested by caterpillars (the larval forms of butterflies, moths, and related insects). In the caterpillar gut, the insecticide lyses the cells of the gut wall. As gut liquid diffuses between the cells, the larvae experience paralysis, and bacterial invasion soon follows (FIGURE 27.7B).

Bacillus thuringiensis toxin (Bt toxin) appears to be harmless to plants and other animals. It is produced by harvesting bacterial cells at the onset of sporulation and drying them into a commercially available dusting powder. The product is useful on butterfly and moth caterpillars. Other strains of the bacterial species produce toxins to beetle and fly larvae, and to mosquitoes. Today, Bt toxin is widely used; more than 32 million hectares (larger than the size of Italy) of land worldwide has been treated with the pesticide.

However, bacterial cells sprayed onto plants can soon wash off, so the protective effect of the insecticide can be limited. DNA technologists realized that long-term protection can be provided by inserting the genes for Bt toxin directly into plants. To date, Bt cotton and Bt corn plants have been developed in the United States. In both cases, the isolated *Bt* gene has been spliced into plant genes using the Ti plasmid of *Agrobacterium tumefaciens* as the carrier (or vector).

Genetic engineering also is being used to develop a single Bt toxin to control many different plant pests. Genetic engineers can engineer a gene encoding the toxic factor along with the factor able to infect different plant pests. Such transformed bacterial cells can be grown in large fermentors to produce the Bt toxin.

Another biotechnological approach to gene-related plant resistance has been used in corn infested with corn borers. Researchers isolated the *B. thuringiensis* gene for toxin production and spliced it into a bacterial species living harmlessly with corn plants. The researchers then forced a colony of gene-altered bacterial cells into corn seeds. When the seeds were sown, the Bt toxin-producing bacterial cells flourished along with the plant, and when an insect ate the plant, it consumed the toxin. This approach is advantageous

(A) (B)

FIGURE 27.7 ***Bacillus thuringiensis*** **and Its Toxic Effects.** **(A)** A false-color transmission electron micrograph of *Bacillus thuringiensis,* showing the development of a spore (orange). (Bar = 0.5 μm.) © George Chapman/Visuals Unlimited. **(B)** A photograph of a dead cabbage butterfly larva (caterpillar) on a leaf treated with Bt toxin. © Holt Studios International Ltd/Alamy Images. »» How did the caterpillar become intoxicated?

because only the insect attacking the plant is subjected to the toxin.

Another useful species is *Bacillus sphaericus,* which kills at least two species of mosquitoes that ingest its poison. To increase the bacterial cell's efficiency, researchers inserted two of its genes into the bacterium *Asticcacaulis excentris* and achieved insecticidal activity against the mosquitoes transmitting malaria, filariasis, and St. Louis encephalitis. Using the gene carrier *A. excentris* is advantageous because it is easier to grow in large quantity; it tolerates sunlight better than *B. sphaericus;* and it floats in water, where mosquitoes feed (the heavier *Bacillus* species sinks because of its spores).

Many bacteriologists are also investigating the insecticidal abilities of the gram-negative rod *Photorhabdus luminescens.* Its toxin, known as Pht, attacks the gut lining of larvae (as Bt does), and is contained in large cytoplasmic crystals; however, its spectrum of activity is wider than for Bt and includes numerous types of caterpillars as well as cockroaches. The Pht genes have been isolated, and efforts are underway to introduce them to plant cells. Normally, *P. luminescens* lives in the intestines of soilborne nematodes. The latter invade insect tissues in the soil, and the bacteria-derived toxin kills the insect.

Viruses also show promise as pest-control devices, partly because they are more selective in their activity than bacterial species. Once released in the field, the viruses spread naturally. It also is possible to harvest infected insects, grind them up, and use them to disseminate the virus to new locations. Among the insects successfully controlled with viruses are the cotton bollworm, cabbage looper, and alfalfa caterpillar.

Researchers also can develop insecticides by using a toxin from the venom of a scorpion. The toxin paralyzes the larvae of moths and other lepidopteran insects. It is attached to a baculovirus, a virus with a high affinity for lepidopteran tissues. Then, the virus is sprayed on lettuce and cotton plants infested with moth larvae. At the conclusion of the field trial, the plot was sprayed with 1% bleach to destroy any remaining viruses.

Viral genes also have been used to protect grapevines. French biotechnologists have incorporated genes from the grape fanleaf virus (GFLV) into champagne grapevines. This virus is transmitted by a nematode and is endemic in the soils of many French regions. It causes malformation of the plant's leaf (the leaf looks like an open fan) and induces the plant to lose chlorophyll and become yellow. To protect the vines, researchers inserted the genes for viral capsids into *A. tumefaciens* and infected the plants with this bacterium. Soon the cells were producing viral capsid proteins, and they became resistant to the virus.

Even a fungus is being employed in the pesticide wars. California researchers have used *Lagenidium giganteum* to protect against crop-damaging mosquitoes in soybeans and rice, and in mosquito-infested nonagricultural settings such as wetlands. The fungus forces its spores into mosquito larvae, which then die in a day or two. Marketed as Laginex®, the fungal preparation has been approved for certain uses by the U.S. Environmental Protection Agency (EPA).

Fungal Organisms Also Are Being Commercially Developed

From the previous sections, it should be obvious that yeasts are involved in many fermentation processes, including alcoholic beverages and baking. Today, yeast cells or their products are of great commercial value.

Yeast cells can be grown in fermentors that are kept aerated. To grow at a maximal rate, plenty of ATP must be made, so aerobic respiration is essential. To keep the cells growing, the energy and carbon source is molasses, which is introduced into the fermentor in small amounts to prevent the yeast cells from switching to fermentation and producing alcohol.

After the growth period, the yeast cells are collected and processed for the commercial market as either dry yeast or as compressed yeast cakes for baking. The product may also be added to commercial wheat or corn flour. Also, because yeast cells are rich in protein and B vitamins, the yeast cells can be dried and marketed in health food stores as a nutritional supplement.

Besides the unicellular fungal yeasts, filamentous fungi also have nutritional value. The most commercial form for human consumption is **mushrooms** (FIGURE 27.8A). In recent years, more and more species of mushrooms are being grown commercially, and now include crimini, shitake, oyster, and portabella. Mushrooms are not grown in fermentors, but rather in special buildings called "mushroom farms" where the temperature and humidity can be closely controlled (FIGURE 27.8A). High humidity and cool temperatures are necessary to trigger mushroom formation.

The beds in which the mushrooms will be grown consist of soil mixed with rich organic matter. The soil is inoculated with a **spawn**, which is a pure culture of the mushroom mycelium. In the bed, the spawn spreads throughout the soil for several weeks. **Casing soil**, a nonnutritious soil layer providing the mycelium with more moisture for mushroom formation, is then added. The appearance of mushrooms on the bed is called a **flush**. Once the flush appears, the mushrooms must be collected while they remain fresh. Commercial farms then package and keep the mushrooms cool for delivery to the market.

(A)

(B)

FIGURE 27.8 **Commercial Mushroom Farming.** **(A)** Many types of mushrooms are grown commercially today. © Jones and Bartlett Publishers. Photographed by Kimberly Potvin. **(B)** A mushroom farm often consists of a rich bed of soil in which the fungal mycelium grows and produces mushrooms in flushes, such as these *Agaricus bisporis*. © Mashkov Yuri, Itar-Tass/Landov. »» Why must the beds be kept in moist and often damp places?

Bioremediation Helps Clean Up Pollution Naturally

Recruiting bacterial species and other microorganisms to break down synthetic waste is an immensely appealing idea. It signals a willingness to work with nature and adapt to its sophisticated sanitation systems, rather than trying to reinvent them. Putting microorganisms to work in this manner is at the crux of **bioremediation**, which is the use of microorganisms to degrade or neutralize contaminants in the soil or water.

■ Night soil: Human excrement collected at night from toilets or cesspools, especially for use as fertilizer.

The concept of bioremediation is not new. In the 1800s, for example, night soil men would, for a small fee, travel from house to house collecting sewage and excrement. After making their rounds, they would scatter their collections on fields, to be broken down by naturally occurring bacterial species in the soil. Although modern waste-disposal systems have replaced the night soil men, a new concern is the plethora of environmental pollutants contaminating the land. Bioremediation seeks to exploit microorganisms to degrade these pollutants.

The advantages of bioremediation were displayed following a major oil spill from the tanker *Exxon Valdez* in 1987 along the Alaska coastline (FIGURE 27.9). Studies have shown that where oil has spilled, the composition of the natural bacterial microbiota rapidly changes to where the oil-degrading species become dominant. Certainly this was the case for the 2010 Deepwater Horizon oil spill in the Gulf of Mexico. Investigating the Microbial World 27 examines these oil-degrading bacterial communities. In the case of the Exxon Valdez spill, technologists only needed to encourage bacterial growth. Thus, technologists "fertilized" the oil-soaked water with nitrogen sources (e.g., urea), phosphorus compounds, and other mineral nutrients to modify the environment and stimulate the growth of naturally occurring microorganisms. Areas treated this way were cleared of oil significantly faster than nonremediated shorelines. Indeed, the oil degraded five times faster when microorganisms were put to work.

FIGURE 27.9 **The Exxon Valdez Oil Spill.** In 1987, the oil tanker Exxon Valdez ran aground on the shoreline of Alaska. Numerous species of naturally-occurring bacteria demonstrated their value by digesting the oil during the ensuing cleanup efforts. This was one of the first large-scale attempts to use microorganisms in bioremediation. Courtesy of the Exxon Valdez Oil Spill Trust Council/NOAA. »» How would these naturally-occurring bacteria be stimulated to rapidly grow to be a significant factor in the cleanup effort?

Bioremediation also can be applied to help eliminate polychlorinated biphenyls (PCBs) from the environment. PCBs were used widely in industrial and electrical machinery before their threat to environmental quality was realized. These inert compounds contain numerous chlorine atoms and chlorine-containing groups, and researchers have identified anaerobic bacterial species able to remove the atoms and their daughter groups, and reduce the compounds to smaller molecules. Aerobic bacterial species now take over and reduce the molecular size still further. Field demonstrations in New York's Hudson River have shown the value of the combination anaerobic–aerobic degradation; where once there was an accumulation of PCBs, now only carbon dioxide, water, and hydrogen chloride remain.

Many years ago, trichloroethylene (TCE) was a much-used cleaning agent and solvent. At the time, scientists did not realize TCE would diffuse through the soil and contaminate underground wells and water reservoirs (aquifers). To combat the problem, scientists have exploited bacterial

Investigating the Microbial World 27

Oil-Eating Microbes

The explosion of the Deepwater Horizon (DH) drilling platform in the Gulf of Mexico in 2010 produced the world's largest accidental release of oil (hydrocarbons) into a body of water. Almost 5 million barrels of light crude were released. By June, 2010, oil had reached the shoreline and beaches in the Gulf where it had some severe ecological consequences.

- **OBSERVATIONS:** Biodegradation of hydrocarbons in the marine environment is primarily mediated by indigenous microbial communities. More than 200 bacterial, algal, and fungal genera making up over 500 species have been identified that can degrade hydrocarbons.

- **QUESTION:** ***In the DH oil spill, what bacterial species are the predominant oil-degrading organisms in contaminated beach ecosystems and how does the bacterial community respond to influx and prolonged oil contamination?***

- **HYPOTHESIS:** The indigenous bacterial communities will increase in numbers in oil-contaminated sands. If so, then by identifying those species in clean and oiled sands, there should be an increase in species population sizes in oiled beach sands.

- **EXPERIMENTAL DESIGN:** Beach sand samples (cores) and trench-excavated sand samples were collected from Pensacola Beach that had been exposed to oil. Control sands came from St. George Island prior to the oil spill.

- **EXPERIMENT 1:** Chemical analysis was used to compare hydrocarbon compositions between oil-contaminated sands and source oil (DH wellhead).

- **EXPERIMENT 2:** Most probable number (MPN) counts and ribosomal RNA (rRNA) gene abundances were used to estimate bacterial populations in experiment 1 samples and compared to clean sand samples.

- **EXPERIMENT 3:** Ribosomal RNA (rRNA) gene sequencing was used to identify the major hydrocarbon-degrading bacterial taxa in the oiled sands.

- **RESULTS:**

EXPERIMENT 1: All beach sand samples contained detectable oil hydrocarbons. However, the sand samples had a 1 to 2 orders of magnitude reduction in the ratio of light molecular weight hydrocarbons (chains of 6 to 16 carbons) to heavy molecular weight hydrocarbons (chains of 16 to 35 carbons) compared to source oil.

EXPERIMENT 2: See table and figure.

TABLE

Enumeration of Oil-Degrading Bacteria—by MPN and rRNA Gene Abundance

Sample Source	MPN Counts (per ml)	rRNA Gene Abundance (copies/g sand)
Oiled sands	2.4×10^{10} cells	0.44×10^7 to 14.2×10^7
"Clean" sands	2.4×10^6 to 9.3×10^6	0.02×10^7 to 1.57×10^7

Data From: Kostka, J. E. *et al.* (2011). Appl. Environ. Microbiol. 77(22): 7962–7974.

Investigating the Microbial World 27 (continued)

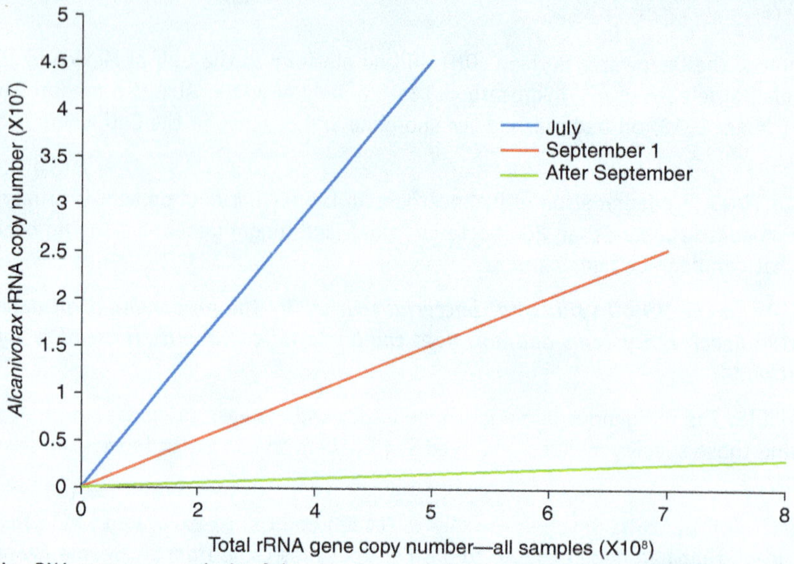

An rRNA gene copy analysis of the abundance of *Alcanivorax* species compared to the abundance of all bacterial species per gram of soil. Data From: Kostka, J. E. *et al.* (2011). *Appl. Environ. Microbiol.* 77(22): 7962–7974.

EXPERIMENT 3: Analysis of bacterial rRNA gene sequences identified 24 bacterial strains from 14 genera. All were known oil-degrading genera. The strains were mostly from the gram-negative Proteobacteria, primarily the Gammaproteobacteria, including *Alcanivorax, Marinobacter, Pseudomonas,* and *Acinetobacter,* and the Alphaproteobacteria, as represented by the Rhodobacteraceae. *Alcanivorax* showed the largest increase in relative abundance among the oiled sand samples.

- **CONCLUSIONS:**

QUESTION 1: ***Was the hypothesis supported? Explain.***

QUESTION 2: ***From experiment 1, what can you conclude about the microbial communities ability to respond to hydrocarbon contamination?***

QUESTION 3: ***What can you conclude from the table concerning the nature of the bacterial community in the "clean" and oiled beach sands? What does the figure suggest concerning Alcanivorax's role in hydrocarbon degradation?***

Answers can be found on the Student Companion Website in **Appendix D.**

Adapted from: Kostka, J. E. *et al.* (2011). *Appl. Environ. Microbiol.* **77**(22): 7962–7974.

species that grow on methane to degrade the TCE. During their metabolism, the bacterial cells produce a methane-digesting enzyme that breaks down TCE. Technologists pump methane and other nutrients into the TCE-contaminated water, and as the bacterial cells grow, they digest the TCE as well as the methane. Once again, the deliberate enhancement of microbial growth yields an environmental cleanup.

Deinococcus radiodurans is a bacterial species able to withstand 3,000 times more radiation than humans. The bacterial cells were found in a tin of irradiation-sterilized ground beef. Researchers are hoping to use this organism in the daunting task of cleaning up thousands of toxic-waste sites where radioactive materials such as plutonium and uranium are present. Genetic engineering methods have produced a *D. radiodurans* strain that degrades ionic mercury compounds common to these sites. The strain uses a gene cluster from *E. coli* to reduce mercury to a less toxic form.

During the 1940s through the 1960s, a major component of weaponry was the explosive compound 2,4,6-trinitrotoluene (TNT). Like other synthetic wastes, this compound has contaminated the soil from residues deposited around weapons plants. Scientists have discovered they can reduce the level of contamination by encouraging bacterial growth with molasses. In a pilot study, researchers mixed water with TNT-laced soil and added molasses at regular intervals. In a matter of weeks, the TNT concentration plummeted.

Plants also can conscript bacterial species for the task of environmental cleanup. Following the Gulf War of 1991, oily devastation remained in much of the Arabian Desert. Within 4 years, however, plant life returned, aided in large measure by *Arthrobacter* species. When researchers dug into oil-soaked desert, they found healthy plant roots surrounded by reservoirs of these oil-degrading gram-negative rods.

For many years, the "haul-and-bury" technique was the prevailing method for disposing of synthetic waste. As the public becomes increasingly intolerant of such an approach, the importance of bioremediation will become more apparent. Technologists are testing microorganisms for their ability to degrade flame-retardants, phenols, chemical warfare agents, and numerous other waste products of industry.

MicroInquiry 27 explores a few more examples with microorganisms.

Industrial Genetic Engineering Continues to Make Advances

In 1973, Herbert Boyer, of the University of California at San Francisco, and Stanley Cohen, of Stanford University, performed the first practical experiment in genetic engineering. Working with *E. coli*, they removed the genes for kanamycin resistance and spliced them into a plasmid already carrying genes for tetracycline resistance. The plasmids then were mixed with *E. coli* cells not resistant to either antibiotic. Finally, the cells were streaked on plates of culture medium containing both kanamycin and tetracycline. The bacterial cells that grew had taken up the plasmids and now were resistant to both antibiotics. The cells had been transformed; they could now do something they could not do before. Feats like these launched the modern era of biotechnology.

Genetic engineering based in plasmid technology has been hailed as the beginning of modern industrial microbiology. **Plasmids** are small, circular, double-stranded DNA molecules that exist apart from the bacterial chromosome (FIGURE 27.10). A single plasmid may contain between 2 and 250 genes. With plasmid technology, a gene of interest is spliced into the plasmid DNA and then the plasmid serves as the vector to carry the foreign gene into the targeted cell. The mechanics of this genetic engineering process are explored in another chapter. Some contemporary products already obtained by plasmid technology include interferon, insulin, and human growth hormone.

When microbiologists first developed plasmid technology, the likely choice for the prototype "bacterial factory" was *E. coli*. Nonpathogenic strains had long been used as test organisms in the laboratory, and the genetics of *E. coli* were well understood. In the 1980s, however, attention shifted to *Bacillus subtilis* and yeasts as host organisms. *B. subtilis* is a gram-positive rod that normally secretes the proteins it makes, while *E. coli* retains them. *Also, B. subtilis* is not regarded as a human pathogen, in contrast to some strains of *E. coli*, nor does it contain endotoxins in its cell wall. Yeast also is a key "factory" due to its traditional role in the fermentation processes and as an organism without disease potential. In fact, *Saccharomyces* species have been reengineered to produce a synthetic vaccine for hepatitis B and to obtain rennin, the enzyme used to make cheese.

The implications of genetic engineering are almost limitless. Biotechnologists are trying to introduce the antibiotic-producing genes of *Streptomyces* species directly into more rapidly growing organisms. They also are trying to amplify the number of plasmids in antibiotic producers to obtain a higher yield of product.

In the 1960s, a "Green Revolution" took place in which genetically modified high-yielding supergrains were exported to poor countries throughout the world to encourage self-sufficiency. However, the program stalled when stocks of essential petroleum fertilizers declined due to high oil prices. Agriculturalists now hope for a second Green Revolution sparked by genetic engineering (FIGURE 27.11). They foresee the day when the

MICROINQUIRY 27

Working with Microorganisms

As we approach the end of this text, you now should have a deep appreciation for the diverse roles microorganisms can have. Certainly, their roles go beyond causing human disease. In this MicroInquiry, let's examine some additional ways microbes are being put to beneficial use. Answers can be found on the Student Companion Website in **Appendix D.**

Oil Exploration Through Microbial Chemistry

For decades, we have been hearing about the impending oil shortage. Indeed, it is becoming harder for oil companies to find deposits of light ("sweet") crude oil, which is the form usually refined. This means about 60% of the world's reserves are a very thick and sticky form called heavy ("sour") crude oil. Much of it remains below ground because it is very difficult to recover and costly to refine into the liquid gasoline you put in your automobile.

27.1a. Outline a microbial way to recover this sour crude oil more reasonably.

Another problem with heavy crude oil is that it contains large amounts of potential contaminants, including sulfur compounds and heavy metals, which could be spread into the atmosphere through refining. The release of these waste products would reverse the trend to cut sulfur emissions in the air. Add to this the high cost of refining such sulfur-rich crude, and you have a dilemma.

27.1b. How can this heavy crude be more economically and safely extracted and refined?

Prospecting with Microbes

The technology of metal extraction has been around for a long time. More than 2,000 years ago, the Romans used iron, copper, lead, gold, silver, and alloys such as bronze and brass for tools, weapons, coins, and jewelry. Unknown to them, microbes had solubilized these metal ores, which the Roman metallurgists had collected from mines in Spain and Wales.

The twentieth century saw an explosion in the production of such high-grade ores, which, like crude oil, are becoming depleted. With this expansion of metal production, besides disfiguring the land, comes the environmental pollution from the smelting process that produces sulfur dioxide in the atmosphere and toxic chemicals in the water and soil.

27.2 Outline a biotechnological scheme to provide for a continued supply of metals.

Superfund and Bioremediation

Historically, and still in use today, are such chemical and physical methods as excavation and incineration, which degrade many of the toxic materials found at Superfund sites. However, the Environmental Protection Agency (EPA) has tested the feasibility of bioremediation at several of these sites.

Creosote Contamination

Between 1973 and 1985, a lumber company in southern Missouri operated a wood treating facility that preserved railroad ties with a creosote/diesel fuel mixture. These operations contaminated the soil at the site with polynuclear aromatic hydrocarbons (PAHs), which were major components of the creosote/diesel mixture. After the facility was shut down and designated as a Superfund site in 1987, the EPA oversaw construction and operation of a land treatment unit to remove the PAH-contaminated soils at the site. EPA studies indicated indigenous microorganisms could digest aromatic hydrocarbons existing in the soil.

27.3 Outline a process whereby excavated contaminated soil could be decontaminated.

Petrochemical Wastes

In Texas, an industrial waste disposal facility contained an estimated 70 million gallons of petrochemical wastes that were disposed of onsite between 1966 and 1971. Contaminants in the lagoon sediments included PAHs, chlorinated organics, and metals.

27.4 Outline a process whereby lagoon sediments could be decontaminated.

genes for nitrogen fixation can be extracted from bacterial organisms like *Rhizobium* and inserted into grain plants such as wheat, rye, and barley. The most optimistic planners look to the future and foresee fertilizers becoming obsolete, plants using microbial toxins to drive off insects, and crops living for weeks without water.

Another dream of biotechnologists is using replacement organisms to interrupt disease cycles in nature. For example, a transgenic snail that resists invasion of *Schistosoma* species could conceivably interrupt the life cycle of the parasite causing schistosomiasis. A slightly different strategy is being employed by DNA technologists who are attempting to produce **transgenic** cotton bollworms by inserting a gene activating a toxin-producing gene in offspring of the bollworm. Work continues to progress on the engineering of a transgenic mosquito unable to harbor and transmit malarial parasites. Released in large numbers, the mosquitoes could dilute or overwhelm native mosquito populations and break the chain of disease transmission.

Superfund: Common name for an environmental law giving the government broad authority to respond directly to releases or threatened releases of hazardous substances endangering public health or the environment.

Transgenic: Referring to an organism containing a stable gene from another organism.

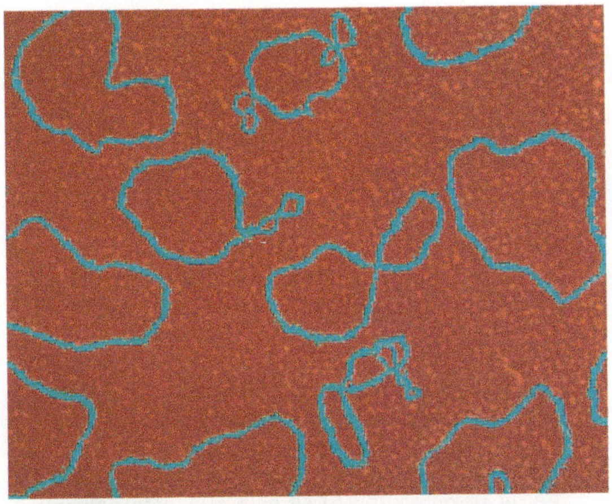

FIGURE 27.10 **Plasmids.** A false-color transmission electron micrograph of bacterial plasmids. Plasmid technology is key to many processes in industrial microbiology and biotechnology. © Dr. Gopal Murti/Visuals Unlimited. »» What types of products may be carried by the information spliced into a plasmid?

FIGURE 27.11 **On the Horizon.** An artist's fanciful conception of a futuristic super plant produced by genetic engineering. Microorganisms or microbial products are key to many of its attributes. »» What concept is depicted by showing this plant growing out of a test tube?

Other research projects in genetic engineering have equally important goals. Research in genetic engineering holds the promise of new strategies for cancer prevention, new diagnostic procedures for microbial diseases and genetic abnormalities, new methods to correct genetic disorders, new hormones, antibiotics, and vaccines, and a generally improved quality of life. The discoveries and insights made possible by genetic engineering have been described as breathtaking. In the future, we can expect startling developments in medicine, agriculture, and the pharmaceutical and chemical industries. Indeed, it is an exciting time to be a microbiologist.

CONCEPT AND REASONING CHECKS 3

a. Why has it been necessary for industry to develop semisynthetic antibiotics, such as many of the penicillins?
b. How are bacterial toxins being produced industrially to kill insect larvae?
c. Summarize the steps involved in mushroom farming.
d. Describe some of the ways bacterial species can be used in bioremediation.
e. Justify the need for industrial microbiology to the advances in biotechnology.

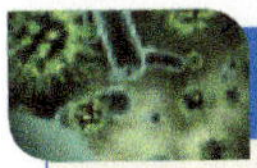

Chapter Challenge C

Okay, we have reached the end of the chapter on industrial microbiology.

QUESTION C:

What additional items have you added to your list of microbial products? If you thought about this at some length, do you believe your list could get even longer? What about the near future? Will biotechnology keep expanding this list? Do microbes rule the world!?

Answers can be found on the Student Companion Website in **Appendix D.**

SUMMARY OF KEY CONCEPTS

27.1 Microorganisms Are Used to Produce Many Industrial Products

1. Among the organic compounds (**secondary metabolites**) synthesized on an industrial scale are organic acids (such as citric, lactic, and gluconic acids), along with various amino acids and vitamins. (Fig. 27.2)
2. Microbes also produce a variety of enzymes such as amylase, pectinase, and protease. Microorganisms produce other key products and save much expense and tedium for the chemist. (Fig. 27.3; Table 27.1)

27.2 Alcoholic Beverages Are Products of Fermentation

3. Yeasts of the genus *Saccharomyces* ferment barley grains to produce beer under anaerobic conditions. The process requires an aging step to develop the full flavor of the product. (Fig. 27.4)
4. Yeasts of the genus *Saccharomyces* ferment grape juice to wine under anaerobic conditions. This process also requires an aging step to develop the full flavor of the product. (Fig. 27.5)
5. To produce spirits, alcohol produced from fermentation is distilled off to produce brandy, rum, whiskey, or neutral spirits.

27.3 Microorganisms Also Produce Many Other Valuble Commercial Products

6. The industrial production of penicillins involves fermentation processes carried out in large fermentors.
7. The spores of *Bacillus thuringiensis* are used in the live form as insecticides for plants and are successfully employed against the caterpillars of various insect pests. In addition, scientists are investigating the use of bacterial insecticides, such as the toxin produced by *Photorhabdus*. Scorpion toxins attached to insect viruses are being developed as potent insecticides.
8. Several fungi have uses as food or food additives. Besides their use in the alcoholic beverage industries, yeasts have commercial value in a dried or cake form in the baking industries. Nutritional yeasts can be a source of protein and B vitamins in the human diet. Mushroom farms have developed the technology to grow many commercial types of wild mushrooms.
9. **Bioremediation** is still another innovative use for industrial microorganisms. In this process, naturally occurring microorganisms are encouraged to grow in a polluted environment and break down pollutants. Bioremediation has been used successfully to degrade the oil in oil spills and to help eliminate polychlorinated biphenyls and trichloroethylene from the environment. Researchers are hoping to use the process in the daunting task of cleaning toxic waste sites that contain radioactive materials.
10. The future of industrial microbiology will center on genetic engineering and plasmid technology. By inserting foreign genes into vector organisms, such as bacterial species and yeast, biotechnologists can produce rare proteins on an industrial scale and provide treatments for such diseases as diabetes, hemophilia, and cancer. Genetic engineers predict plants will be developed with inborn resistance to disease, animals will be engineered to produce human proteins, and novel treatments will be designed to cure or prevent specific illnesses. Medicine, agriculture, and industry eagerly anticipate the fruits of modern biotechnology that are made possible with microorganisms.

CHAPTER SELF-TEST

For **STEPS A–D**, answers to even-numbered questions and problems can be found in **Appendix C** on the Student Companion Website at **http://microbiology.jbpub.com/10e**. In addition, the site features eLearning, an online review area that provides quizzes and other tools to help you study for your class. You can also follow useful links for in-depth information, read more MicroFocus stories, or just find out the latest microbiology news.

STEP A: REVIEW OF FACTS AND TERMS

Multiple Choice

Read each question carefully, then select the ***one*** answer that best fits the question or statement.

1. Which one of the following is NOT true of a primary metabolite?
   A. It is directly involved with normal growth.
   B. It is normally produced during the stationary growth phase.
   C. Alcohol is a primary metabolite.
   D. It is a product essential to the survival of the microbe.
2. Industrially produced streptokinase is a/an _____ used to _____.
   A. enzyme; dissolve blood clots
   B. vitamin; prevent pernicious anemia
   C. enzyme; facilitate the absorption of fluids
   D. enzyme; digest glucose
3. The brewing of a lager requires _____.
   A. a bottom yeast
   B. *Saccharomyces cerevisiae*
   C. a top yeast
   D. wild yeasts
4. In wine making, wild yeasts often _____.
   A. are killed with sulfur dioxide
   B. produce undesirable wine qualities
   C. are found on the grapes
   D. All the above (**A–C**) are correct.
5. A distilled spirit that is 60 proof contains _____.
   A. 15% alcohol
   B. 30% water
   C. 30% alcohol
   D. 60% alcohol

6. Most antibiotics are the products of _____ species.
   A. *Streptomyces*
   B. *Penicillium*
   C. *Streptococcus*
   D. *Bacillus*
7. The toxin produced by *Bacillus thuringiensis* is _____.
   A. a protein
   B. an acid
   C. a sugar
   D. a cell wall component
8. A spawn is a _____.
   A. complex set of nutrients
   B. cluster of mushrooms
   C. bed in which mushrooms grow
   D. mushroom mycelium
9. What organism is being tested for use in the bioremediation of toxic waste sites where radioactive materials are present?
   A. Yeasts
   B. *Bacillus*
   C. *Pseudomonas*
   D. *Deinococcus*
10. For plasmid technology, *Bacillus* _____ may be the "bacterial factory" because it produces no _____.
   A. *cereus;* cell wall
   B. *subtilis;* endotoxins
   C. *thuringiensis;* exotoxins
   D. *subtilis;* cell wall

Matching

Many different microorganisms find value in industrial microbiology. To test your knowledge of these organisms, match the microorganism on the right to the characteristic on the left by placing the correct letter in the space. A letter may be used once, more than once, or not at all.

Characteristic

_____ **11.** Used to ferment grapes.
_____ **12.** Used in riboflavin production.
_____ **13.** Produces many antibiotics.
_____ **14.** Enhances grape sugar content.
_____ **15.** Used in lysine production.
_____ **16.** Biological insecticide.
_____ **17.** Source of musk oil.
_____ **18.** Used for citric acid production.
_____ **19.** Fungal source of amylase.
_____ **20.** Produces ergot poisons.
_____ **21.** Used in top fermentation of beer.
_____ **22.** Yeast for making pilsner beer.
_____ **23.** Able to degrade oil.
_____ **24.** Used for genetic engineering.
_____ **25.** Degrades ionic mercury compounds.

Microorganism

A. *Acetobacter aceti*
B. *Ashbya gossypii*
C. *Aspergillus niger*
D. *Aspergillus oryzae*
E. *Bacillus thuringiensis*
F. *Bacillus subtilis*
G. *Botrytis cinerea*
H. *Claviceps purpurea*
I. *Deinococcus radiodurans*
J. *Enterobacter aerogenes*
K. *Pseudomonas* species
L. *Rhizopus nigricans*
M. *Saccharomyces carlsbergensis*
N. *Saccharomyces cerevisiae*
O. *Saccharomyces ellipsoideus*
P. *Streptomyces* species
Q. *Ustilago zeae*

STEP B: CONCEPT REVIEW

26. Distinguish between **primary** and **secondary metabolites.** (**Key Concept 1**)
27. Summarize the types of enzymes produced by the industrial fermentation of microorganisms for commercial and medical needs. (**Key Concept 1**)
28. Construct a concept map for beer production. (**Key Concept 2**)
29. Describe the steps in the production, fermentation, and aging of wine. (**Key Concept 2**)
30. Compare and contrast the production of distilled spirits from wine making. (**Key Concept 2**)
31. Explain why antibiotics produced through industrial fermentation represent secondary metabolites. (**Key Concept 3**)
32. Evaluate the role of microorganisms to the industrial production of insecticides. (**Key Concept 3**)
33. Identify the usefulness of fungi to food microbiology and as a food supplement. (**Key Concept 3**)
34. Explain how bioremediation can be accelerated using microorganisms. (**Key Concept 3**)
35. Describe how plasmid technology is advancing biotechnology. (**Key Concept 3**)

STEP C: APPLICATIONS AND PROBLEMS

36. Many beer companies have developed strains of yeasts that break down more of the carbohydrate in barley malt than traditional yeasts. What do you think their product is called?

37. The discovery of the extremely high resistance of *Deinococcus radiodurans* to radioactivity has prompted hopes that this organism can be used in bioremediation. Suppose you were to go on a hunt for novel organisms that could be used for other environmental cleanups. Where might you look?

38. A product called Dipel dust contains *Bacillus thuringiensis*. It is used in the Northeast for destruction of the gypsy moth caterpillar. What dangers might result from extensive application of this product?

39. A friend who also has taken a class in microbiology wonders if it would be possible to engineer certain bacterial species to produce antibiotics and then give these species to diseased people to serve as antibiotic producers within the body. How would you respond to your friend's idea?

STEP D: QUESTIONS FOR THOUGHT AND DISCUSSION

40. The poet John Donne once wrote: "No man is an island, entirely of itself." This maxim applies not only to humans, but to all living things in the natural world. What are some roles the microorganisms play in the interrelationships among living things?

41. When the Mayflower set sail for the New World, its intended destination was Virginia. Instead, it landed at Plymouth, Massachusetts, because, as one diarist put it, "We could not now take time for further search or consideration, our victuals being much spent, especially our beer." What do these last few words tell you about the Pilgrims?

42. In several places in this text, we have noted how apparently harmless organisms have been discovered later to be dangerous in humans. Yeasts have been consumed in breads and alcoholic beverages for centuries, and still no pathogenic signs have been observed. Can you postulate why?

43. Certain bacterial species produce many thousands of times their required amount of specific vitamins. Some biologists suggest that this makes little sense because the excess is wasted. Can you suggest a reason for this apparent overproduction in nature?

44. How many times in the last 24 hours have you had the opportunity to use or consume the industrial product of a microorganism?